3

INTRINSICALLY DISORDERED PROTEINS (IDPs)

STRUCTURAL CHARACTERIZATION, THERAPEUTIC APPLICATIONS AND FUTURE DIRECTIONS

PROTEIN BIOCHEMISTRY, SYNTHESIS, STRUCTURE AND CELLULAR FUNCTIONS

Additional books in this series can be found on Nova's website
under the Series tab.

Additional e-books in this series can be found on Nova's website
under the e-book tab.

PROTEIN BIOCHEMISTRY, SYNTHESIS, STRUCTURE
AND CELLULAR FUNCTIONS

INTRINSICALLY DISORDERED PROTEINS (IDPS)

STRUCTURAL CHARACTERIZATION, THERAPEUTIC APPLICATIONS AND FUTURE DIRECTIONS

VIOLET WEBER
EDITOR

nova publishers
New York

NOTICE TO THE READER

Library of Congress Cataloging-in-Publication Data

ISBN: 978-1-63484-407-9
Library of Congress Control Number: 2015959367

Published by Nova Science Publishers, Inc. † New York

CONTENTS

PREFACE

Intrinsically disordered proteins (IDPs) are biomolecules that do not have a definite 3D structure; their role in the biochemical network of a cell relates to their ability to switch rapidly among different secondary and tertiary structures. The emergence of IDPs has challenged the classical protein structure-function paradigm. IDPs play an important role in cellular regulation, signaling and control in health and disease. However, the unusual biophysics of these proteins makes structural characterization of IDPs and their complexes not only challenging but often resulting in opposite conclusions. This book studies the structural characterization and future directions of IDPs. The first chapter focuses on the DNA-binding IDPs and how intrinsic disorder affects their function. The second chapter discusses the use of a computer atomistic simulation for the structural analysis of IDPs. The final chapter, examines two long-standing contradictions concerning dimerization and membrane-binding activities of IDPs; provides an explanation of these discrepancies; and demonstrates how the resolution of these critical issues in the field results in the expanded understanding of cell function with multiple applications in biology and medicine.

In: Intrinsically Disordered Proteins (IDPs)   ISBN: 978-1-63484-407-9
Editor: Violet Weber   © 2016 Nova Science Publishers, Inc.

Chapter 1

INTRINSIC DISORDER IN DNA-BINDING PROTEINS

Alicia Roque, Inma Ponte and Pedro Suau

Departamento de Bioquímica y Biología Molecular,
Facultad de Biociencias, Universidad Autónoma de Barcelona, Spain

ABSTRACT

Intrinsically disordered proteins (IDPs) or intrinsically disordered regions (IDPRs) do not have unique 3D structures under physiological conditions. They share specific sequence features such as low overall hydrophobicity and high net charge. They also have a distinctive conformational behavior, with large hydrodynamic volumes, low contents of ordered secondary structure, and high structural heterogeneity. Despite their flexibility, some IDPs/IDPRs may undergo disorder to order transitions in the presence of natural ligands. IDPs carry out numerous biological functions, including those associated with signaling, transcription regulation, DNA condensation, cell division, and cellular differentiation. Bioinformatic prediction tools revealed that as many as 50% of eukaryotic proteins are likely to contain functionally important long disordered regions. The highest number of predicted IDPs or IDPRs corresponds to DNA-binding proteins. However, the number of

* Corresponding author: Pedro Suau, Departamento de Bioquímica y Biología Molecular, Facultad de Biociencias, Universidad Autónoma de Barcelona, 08193 Bellaterra, Barcelona, Spain. Phone: 34 935811391, Mobile: 34639342418, Fax: 34 935811264, Email: pere.suau@uab.es.

experimentally confirmed IDP/IDPRs included in the Disprot database is quite smaller, comprising only 73 proteins. This functional subclass is composed mainly of transcription factors, DNA metabolism proteins and chromosomal architectural proteins like histones and high mobility group proteins (HMGs). In this chapter we will focus on the DNA-binding IDPs and how intrinsic disorder affects their function.

INTRODUCTION

Many protein regions and even entire proteins lack a stable tertiary and/or secondary structure in solution, while possessing numerous biological functions. They are known as intrinsically disordered proteins (IDPs) or intrinsically disordered protein regions (IDPRs) (Uversky, 2012). IDPs carry out numerous biological functions, especially those associated with signaling, transcriptional regulation, DNA condensation, cell division, and cellular differentiation (Dunker et al., 2015).

Intrinsically disordered proteins differ from globular proteins in several aspects including amino acid composition. They have low amounts of order-promoting residues like I, V, L, F, Y and W, which would normally form the hydrophobic core of a folded globular protein. On the other hand, IDPs/IDRs are enriched in disorder-promoting amino acids like A, R, G, Q, S, P, E and K. IDPs and globular proteins are also different in sequence complexity, hydrophobicity, net charge, conformational flexibility and type and rate of amino acid substitutions over evolutionary time. IDPs are also characterized by their large hydrodynamic volumes and low contents of ordered secondary structure (Uversky, 2012).

Conformational plasticity in IDPs can have many advantages including increased interaction surface, increased interaction speed, determined by its higher capture radius, the ability to bind many interaction partners by different conformations, the ability to fold partially or completely upon binding, the ability to form fuzzy complexes and the high accessibility of sites for post-translational modifications (PTMs), which facilitate their regulation (Uversky, 2013).

Intrinsic disorder is quite common in eukaryotes. As much as 50% of eukaryotic proteins are likely to contain functionally important long disordered regions. When classified by ligand, DNA-binding proteins are the largest group of predicted IDPs (Xie et al., 2007). This group is mainly composed of nuclear architectural proteins including histones and non-histone proteins, transcription factors and proteins involved in DNA metabolism. In this chapter, examples of IDPs belonging to those different kinds of DNA-binding proteins are reviewed.

1. CHROMATIN ARCHITECTURAL PROTEINS

Eukaryotic DNA is packed into chromatin. This nucleoprotein complex is made up of repeating nucleosome units. Nucleosomes consist of 147 bp of DNA wrapped around an octamer of core histones (two copies each of H2A, H2B, H3 and H4) that are connected to each other by short stretches of linker DNA. Another type of histone, known as H1, binds to the nucleosome and linker DNA.

The accessibility of eukaryotic DNA is dependent upon the hierarchical level of chromatin organization. Chromatin condensation depends on intra- and inter-nucleosome interactions, as well as on the influence of non-histone chromatin architectural proteins (McBryant et al., 2006). DNA-binding proteins such us protamines, high mobility group proteins (HMGs), heterochromatin protein 1 (HP1) and methyl-binding protein 2, (MeCP2) are classified as non-histone chromatin architectural proteins and play a role in the regulation of chromatin higher-order structure.

Many chromatin architectural proteins contain intrinsically disordered regions lacking regular secondary structure, suggesting that this may be a common mechanism by which the cell manages and maintains chromatin dynamics (McBryant et al., 2006). In this chapter we will review histone and non-histone chromatin architectural proteins containing functionally relevant intrinsically disordered regions.

1.1. H1 Linker Histones

H1 linker histones bind to linker DNA regions on the surface of the nucleosome. H1 is present at about one molecule per nucleosome in metazoans. H1 is thought to be primarily responsible for the condensation of the thick chromatin fiber. Histone H1 could have a regulatory role in transcription through the modulation of chromatin higher-order structure. H1 has been considered as a general transcriptional repressor because it contributes to chromatin condensation, which limits the access of the transcriptional machinery to DNA. However, H1 may mediate transcription at a more specific level, participating in complexes that either activate or repress specific genes (Zlatanova et al., 2000; Harshman et al., 2013).

H1 has multiple isoforms. In mammals, seven somatic subtypes (designated H1.0-H1.5 and H1.10), three male germ-line-specific subtypes, (H1.6, H1.7 and H1.9) and an oocyte-specific subtype (H1.8) have been identified (Talbert et al., 2012). H1 subtypes have different affinities for DNA and chromatin and affect the nucleosome repeat length (NRL) (Orrego et al., 2007; Sancho et al; 2008; Th'ng et al., 2005).

Protease digestion and NMR analysis showed that linker histones contain three distinct domains: a short amino-terminal domain (NTD) (20-35 amino acids), a central globular domain (GD) (~80 amino acids), and a long carboxy-terminal domain (CTD) (~100 amino acids) (Chapman et al., 1976) (Figure 1A). The most conserved region of histone H1 is the GD. The N- and C-terminal domains are, in general, more variable. The H1 subtypes show a wide variety of substitution rates, which is in favor of their functional differentiation (Ponte et al., 1998).

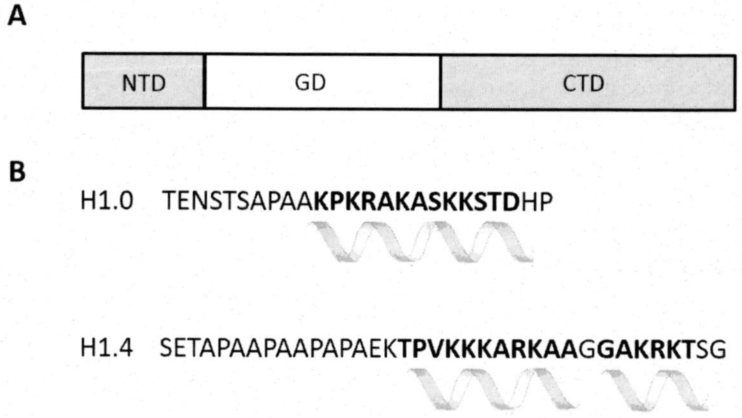

A

NTD	GD	CTD

B

H1.0   TENSTSAPAA**KPKRAKASKKSTD**HP

H1.4   SETAPAAPAAPAPAEK**TPVKKKARKAA**GG**AKRKT**SG

Figure 1. Intrinsic disorder in histone H1. A. Structure of histone H1. H1 is composed of three structural domains: the N-terminal domain (NTD), the globular domain (GD) and the C-terminal domain (CTD). In light gray, intrinsically disordered regions. B. Secondary structure in the NTD of mouse H1.0 (Uniprot, P10922) and H1.4 (Uniprot, P43274) determined by NMR (Vila et al., 2001a; Vila et al, 2002). In bold the residues in α-helix.

The classical tripartite structure of H1 histones is found in higher eukaryotes. Some protists, such as *Tetrahymena thermophila*, have proteins that lack of the stably folded GD and are similar to the C-terminal domain

(Hayashi et al., 1987). On the other hand, yeast has an H1-like nuclear protein with the classical three-domain structure plus an extra globular domain (Ali et al., 2004). This additional globular domain has the properties of a molten globule.

1.1.1. Secondary Structure of the Amino-Terminal Domain of Histone H1

The N-terminal domain (NTD) of histone H1 is the shortest domain of the protein with a length between 20 and 35 residues, depending on the subtype. The N-terminus can be divided into two sub-regions (Bohm and Mitchel, 1985). The distal part of the NTD is enriched in hydrophobic residues, whereas a highly basic region is found close to the globular domain (Böhm and Mitchel, 1985). The NTD is mainly unstructured in solution, but NMR studies of peptides derived from the NTD of the subtypes H1.0 and H1.4 (equivalent to H1e) show that the basic cluster can adopt α-helical structure in the presence of trifluoroethanol (TFE), which is a stabilizer of secondary structure (Vila et al., 2001a; Vila et al., 2002). The NTD of H1.4 contains two α-helical regions. The helical elements are connected by a Gly–Gly motif. Structure calculations show that the Gly–Gly motif behaves as a flexible linker between the helical regions. The wide range of relative orientations of the helical axes allowed by the Gly–Gly motif may facilitate the tracking of the phosphate backbone by the helical elements or the simultaneous binding of two nonconsecutive DNA segments in chromatin (Figure 1B) (Vila et al., 2002).

The study of two peptides belonging to the NTD of H1.0 revealed the presence of a single α-helical element, spanning from K11 to D23, with the first three basic residues, K11, K13 and R14, on one side of the helix and the following three basic residues, K16, K19 and K20, on the other side (Figure 1B). The analysis of the complexes of these peptides with DNA by Fourier Transform Infrared Spectroscopy (FTIR) showed the induction of α-helix with percentages in agreement with the helical propensities observed in trifluoroethanol (TFE). This result indicates that the basic region of the NTD is an intrinsically disordered region with coupled binding and folding (Vila et al., 2001a).

The NTD has been shown to be non-essential for the formation of higher-order chromatin structures, but its deletion appears to reduce the affinity of H1 for chromatin (Öberg and Belikov, 2012). FRAP analysis of chimeric H1, where the NTD of H1.0 and H1.2 had been swapped, suggests that the NTD contributes to the differential binding affinities of the subtypes (Vyas et al.,

2012). The induction of α-helix within the NTD could be essential for its proper binding to chromatin. PTMs and protein-protein interactions could also play a role in the modulation of the NTD structure.

1.1.2. *Secondary Structure of the Carboxy-Terminal Domain of Histone H1*

The CTD has approximately 100 amino acids and 30-50 net positive charges contributed mostly by lysines. The CTD is the primary determinant of H1 binding to chromatin *in vivo* (Hendzel et al., 2004). Several studies indicate that the ability of histone H1 to stabilize chromatin folding resides in the CTD (Allan et al., 1980; Lu and Hansen, 2004; Allan et al., 1986).

The H1 terminal domains have been traditionally referred to as tails because they have little structure in aqueous solution. However, the CTD acquires a proportion of α-helical structure in the presence of trifluoroethanol, alkaline pH or high salt (Clark, 1988). The conformational features of a peptide belonging to the CTD of H1.0 were studied by high-resolution NMR (Vila et al., 2001b). The peptide of 23 amino acids is adjacent to the globular domain and contains the longest proline-free fragment of the CTD. In aqueous solution, the peptide was mostly unstructured, but in the presence of trifluoroethanol it acquired a substantial amount of α-helical structure. The helical region, spanning 17 residues, has a TPKK motif in a turn conformation at the C-terminal end.

The low resolution structure of the entire CTD in solution and bound to DNA was studied by FTIR (Roque et al., 2005). The CTD has little structure in aqueous solution, but becomes extensively folded upon interaction with DNA. The secondary structure elements present in the bound CTD include α-helix (24%), β-structure (25%), open loops (17%) and turns (33%). The CTD thus behaves as an intrinsically disordered protein with coupled binding and folding. These results are supported by Fluorescence Resonance Energy Transfer (FRET) experiments. Labeling with Cy3 and Cy5 of H1 on either end of the CTD showed that the CTD had an extended structure in solution. Interaction with nucleosomes elicited a drastic increase in FRET, indicating the reduction in the distance between the ends of the CTD, following the folding of the domain (Caterino et al., 2011; Caterino and Hayes, 2011; Fang et al., 2012). The estimated FRET efficiency indicates a reduction of the mean distance between the ends of the CTD from ~9 nm in the free protein to ~5.2 nm when bound to nucleosomes.

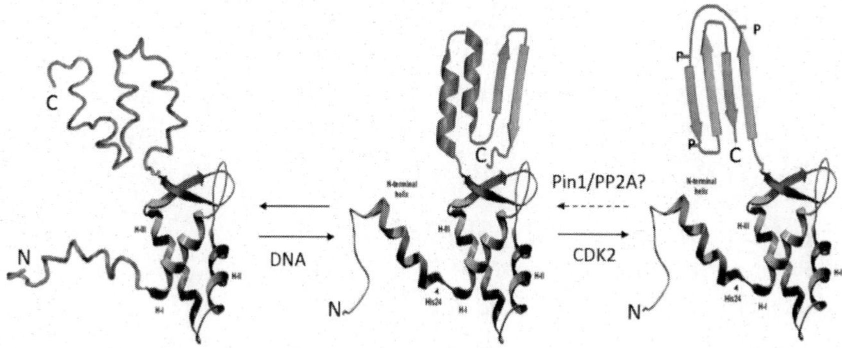

Figure 2. Schematic model of the folding of mouse histone H1.0. Both intrinsically disordered domains, the NTD and the CTD fold upon binding to DNA (Vila et al., 2002; Roque et al., 2005). N, N-terminus; C, C-terminus; P, phosphate group. Phosphorylation by CDK2 of the CTD causes a conformational change in this domain. The dashed arrow represents the possible involvement of Pin1 and PP2A phosphatase in the reversal of the conformational changes induced by CDK2 phosphorylation.

1.1.3. Role of Charge Neutralization in the Folding of the CTD of Histone H1

The interaction of the CTD with DNA is basically electrostatic. Folding of the CTD upon DNA interaction may follow extensive compensation of the Lys positive charge by the DNA phosphates. The effects of electrostatic shielding of the DNA phosphates on DNA condensation and bending have been thoroughly studied (Manning, 2003). The reciprocal effects on protein folding of the neutralization of basic residues by DNA phosphates have received less attention, particularly in the case of basic proteins with intrinsic disorder. The folding of the CTD at alkaline pH was shown for the first time by Clark et al., (Clark, 1988). Recently, FTIR was used to show that charge suppression at alkaline pH induced extensive folding of the CTD, with proportions of secondary structure motifs similar to those observed in the complexes with DNA (Roque et al., 2009).

The question arises as to the precise role of charge neutralization in the folding of the CTD. The electrostatic repulsion between basic residues is minimized at alkaline pH, and this could contribute to the folding of the polypeptide chain. The Lys side-chain does have a significant hydrophobic character due to its four methylene groups. Deprotonation thus implies an increase in the mean hydrophobicity of the CTD that leads to folding. The effect should be significant given the high Lys content of the CTD (~40%). The similarity of the conformational effects of charge suppression at alkaline pH and charge compensation in the DNA complexes suggests that folding of the CTD upon interaction with DNA could be driven by increased hydrophobicity of the CTD upon electrostatic neutralization of the Lys charge by DNA phosphates. Partial neutralization would generate a large variety of partially folded conformations.

1.1.4. The CTD As a Molten Globule

Proteins are usually studied in dilute solution; however, in the cellular environment, macromolecules and small-molecule solutes are present at high concentrations, so that significant fraction of the intracellular space is not available to other macromolecules. The excluded-volume effects are predicted to favor the adoption of compact, as opposed to expanded, macromolecular conformations, resulting in a reduction of the total excluded volume. IR spectroscopy was used to estimate the proportions of secondary structure motifs of the CTD of H1.0 and H1.6 in the presence of the crowding agents Ficoll 70 and polyethylene glycol (PEG) 6000 (Roque et al., 2007). Crowding appeared to be highly effective in promoting the folding of the CTD. In the presence of 30% PEG or Ficoll, the proportions of secondary structure motifs; i.e., α-helix, β-structure, turns and open loops, were similar to those of the DNA-bound domain. Kratky plots of the small-angle X-ray scattering showed that in crowding agents the CTD has the compaction of a globular state. Despite the importance of charge neutralization in the folding of the CTD, the fact that the presence of crowding agents at neutral pH induces a molten globule state in the CTD with native-like secondary structure and compaction does not support the presence of a large unfavorable electrostatic contribution to the folding free energy.

The condensation of the CTD into a native-like structure may increase the rate of the transition toward the native-bound state. The folding of the CTD in crowding conditions may also facilitate diffusion of H1 inside cell nuclei. Improved diffusion could have functional consequences because H1 molecules exchange *in vivo* between chromatin binding sites through soluble intermediates (Lever et al., 2000).

1.1.5. Intrinsic Disorder and Binding of the CTD

Some models based on FRAP analysis assume that binding of H1 to DNA/nucleosomes in chromatin is initiated by nonspecific electrostatic interactions of the CTD with DNA phosphates that would induce the acquisition of secondary and tertiary structure and increased affinity. Folding upon binding may proceed either from basically unordered states or from premolten or molten globule states, depending on the local degree of crowding experienced by the CTD. Were it to be in a molten globule state, folding upon binding would simply consist in the structural "tightening" of the molten globule. Binding of the CTD to unspecific sites, in nucleosomal and linker DNA, may lead to partial charge neutralization and thus to partial folding, lower affinity and shorter residence times. Intrinsic disorder could thus accelerate the recognition of specific binding sites by the GD by shortening the residence time of H1 in non-specific sites. Interaction of chromatin with histone H1 containing swapped N- and C-terminal domains shows that the ability to properly fold the chromatin fiber is impaired. Furthermore, FRAP analysis showed a decreased time of residence. These results confirm that in order to fulfill its functions, both the GD and the CTD must bind properly (Hutchinson et al., 2015).

Folding of the highly basic CTD is dependent on the neutralization of the ε-amino groups of the lysine residues by the DNA phosphates. The interesting point is that the amount of secondary structure is proportional to the degree of charge neutralization. Intrinsic disorder would thus allow the CTD to adopt a continuum of structures going from those with only residual structure to those fully structured, passing through multiple intermediate states. The presence of subdomains in the CTD may contribute to the variety of partially folded conformations. A possible advantage of an intrinsically disordered CTD would thus be to kinetically favor the recognition of H1 binding sites.

1.1.6. Phosphorylation of the CTD of Histone H1

H1 histones are targets of several post-translational modifications, including phosphorylation, acetylation, methylation, formylation, ubiquitination, deamidation and citrullination. The phenotypic roles of histone H1 may be determined by complementary and overlapping effects of stoichiometry, subtype composition and post-translational modifications. The main post-translational modification of histone H1 is phosphorylation of the consensus sequences (S/T)-P-X-(K/R) by cyclin dependent kinases (CDKs). In mammalian subtypes, these sequences are located mostly in the CTD. The highest number of phosphorylated sites is found in mitosis, where chromatin is maximally condensed. During interphase, H1 subtypes are present as a mixture of unphosphorylated and low-phosphorylated species, with a proportion of 35%–75% of unphosphorylated forms, according to the particular subtype and cell-line and the moment of the cell cycle. It is not clear how H1 phosphorylation by CDK2 affects chromatin condensation during interphase and mitosis. A number of studies indicate that interphase phosphorylation is involved in chromatin relaxation (Roth and Allis, 1992; Green et al., 2011; Talasz et al., 2009; Zheng et al., 2010; Kostova et al., 2013; Th'ng et al., 1994); however, in metaphase chromosomes, H1 is hyperphosphorylated. These features support the view that partial and full phosphorylation are functionally distinct.

The effects of phosphorylation on the secondary structure of the CTD were studied by FTIR (Roque et al., 2008). Full phosphorylation (3 or 4 phosphate groups, depending on the subtype) of the DNA-bound CTD brings about a large structural change consisting in a significant increase in β-structure accompanied by a decrease in α-helix. Partial phosphorylation induces, in general, a large proportion of undefined structure and a decrease in α-helix, without a significant increase in β-structure. The basic conformational tendencies of the fully phosphorylated domain when bound to DNA; i.e., more β-structure and less α-helix were also observed at alkaline pH, confirming that phosphorylation changes the conformational trends of the CTD [Roque et al., 2009; Roque et al., 2012].

The effects of phosphorylation on the affinity of the CTD for the DNA were moderate and depended on the number of phosphate groups. Partial phosphorylation drastically reduced the aggregation of DNA fragments by the CTD, but full phosphorylation restored to a large extent the aggregation capacity of the unphosphorylated domain. These results support the involvement of H1 hyperphosphorylation in metaphase chromatin condensation and of H1 partial phosphorylation in interphase chromatin relaxation.

Linker histones in chicken erythrocyte soluble chromatin were partially phosphorylated with CDK2. FTIR difference spectroscopy showed a gradual increase in β-structure in the phosphorylated samples, concomitant to a decrease in α-helix/turns, with increasing linker histone phosphorylation. This conformational change could act as the first step in the phosphorylation-induced effects on chromatin condensation (Lopez et al., 2015).

The effects of phosphorylation on chromatin condensation and aggregation have been explained by the decrease of the positive net charge that would weaken the interaction of H1 with chromatin (Dou and Gorovsky, 2000). However, the structural change brought about by phosphorylation, characterized by a strong induction of β-structure, may have a more significant effect, mainly if we consider the characteristic distribution of the basic residues in the CTD. About 70% of the Lys residues of the CTD are present in doublets that would project each of the lysine residues in opposite directions in the plane of the β-sheet (Roque et al., 2009). This arrangement may lower the local level of DNA charge neutralization and, thus, the H1 binding affinity for chromatin. This would explain the general role of partial phosphorylation in promoting chromatin relaxation (Kostova et al., 2013).

The study of the interaction of Pin1, a peptidyl-prolyl isomerase, and H1 provided additional evidence of the conformational change of H1 following phosphorylation (Raghuram et al., 2013). Pin1 catalyzes the cis-trans isomerization of the (S/T)-P bond and specifically interacts with phosphorylated H1. Its catalytic activity is necessary to achieve the normal rate of H1 dephosphorylation by PP2A phosphatase, which specifically dephosphorylates the trans-isomer of the (S/T)-P bond (Zhou et al., 2000). The role of Pin1 would be to catalyze the isomerization of the cis-isomer, apparently present in phosphorylated H1, to the trans-isomer, which can be dephosphorylated.

The physiological role of Pin1 was demonstrated by FRAP analysis, which showed that H1.1 and H1.5 had higher residence time in chromatin in the presence of Pin1, due to Pin1-mediated dephosphorylation of serine residues.

Figure 2 shows a schematic model based on the percentages of secondary structure motifs obtained by FTIR, structural predictions and FRET experiments (Roque et al., 2005; Roque et al., 2008; Caterino et al., 2011; Caterino and Hayes, 2011; Fang et al., 2012 and Roque et al., 2009). This model shows the folding of the unphosphorylated CTD bound to DNA and incorporates the increase in β-structure induced by phosphorylation. Functional evidence suggests that this conformational change may be reversed by the combined actions of Pin1 and PP2A phosphatase (Raghuram et al., 2013).

Phosphorylation of the DNA-bound CTD by cyclin-dependent kinases appears as a main conformational modulator, affecting the proportions of α-helix, turns, unordered structure and β-structure. The effects of phosphorylation may be thus mediated by specific structural changes and not be a simple effect of a decrease in the net charge. In summary, intrinsic disorder together with PTMs confers the CTD a remarkable structural flexibility.

1.2. Core Histones

Core histones are highly conserved small basic proteins. At the C-terminus they possess a well-structured histone fold domain which is involved in histone-histone and histone-DNA interactions (Figure 3A). The histone fold domain consists of three α-helical elements. The N-terminal domain is intrinsically disordered, extends out of the nucleosome and is essential for chromatin condensation and oligomerization (Carruthers and Hansen, 2000; Kan et al., 2007; McBryant et al., 2009). The NTDs of core histones are rich in disorder promoting residues. All core histone NTDs contain lysine, serine and glycine, but differ in the proportion of other residues like proline, arginine and alanine (Hansen et al., 2006).

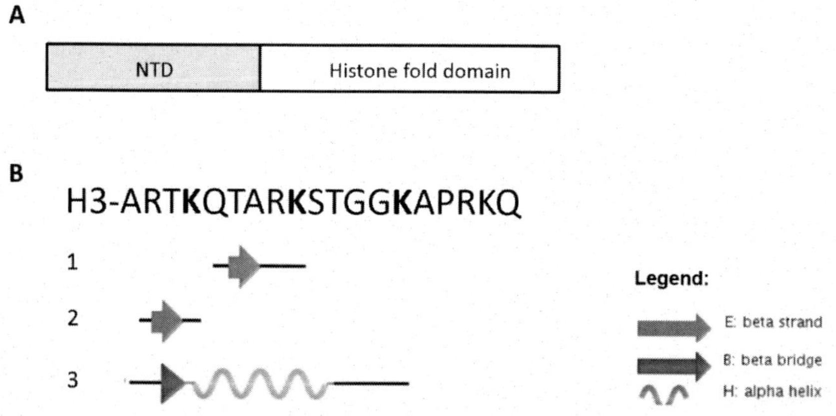

Figure 3. Structural domains and folding of core histones. A. Domains of core histones. NTD, N-terminal domain. In light gray, intrinsically disordered regions B. Secondary structure motifs of H3 post-translationally modified in complex with several chromatin reader modules. The sequence corresponds to the first 20 aminoacids of human H3 (Uniprot, P68431). B1, structure found in an H3 peptide containing H3K9me3 in complex with HP1 (PDB, 1KNE). B2, structure found in an H3 peptide containing H3K4me3 in complex with the tudor-like domain of Spindlin1 (PDB, 4H75). B3, structure found in an H3 peptide containing H3K14ac in complex with MOZ double PHD finger (PDB, 4LLB). In bold the residues post-translationally modified in the crystal structures.

1.2.1. Folding of the Core Histones NTD

Analysis of the primary sequence of core histones with bioinformatic tools predicts helical propensities in their NTD. Experimental data corroborate these predictions. Circular dichroism studies comparing the intact nucleosome core particle with the half-proteolyzed particle with both H3 and H4 N-termini selectively removed, and the fully proteolyzed particle, devoid of all the core histone N termini suggested that the NTD of H3 and H4 acquired significant proportions of α-helix when bound to DNA (Banères et al., 1997). In support of these results, differences between the helical content of the nucleosome in solution and the nucleosome core particle were found by circular dichroism, which accounted for a 15–20% increase in helical content upon binding to DNA, presumably due to the folding of the NTD (Wang et al., 2000).

The NTD of core histones is rich in post-translational modifications. The interplay of histone PTMs constitutes the "histone code," which plays an important role in the regulation of diverse cellular processes, including chromatin condensation, gene expression, cell differentiation, and development.

If charge neutralization of the core histone NTDs upon interaction with DNA is responsible for their α-helical conformation, then it would be expected that lysine acetylation, which decreases the positive net charge in the NTD would favor this conformation. Early studies indicated that histone acetylation increases the overall α-helical content. These findings are supported by circular dichroism analysis of acetylated H4 in solution and in the nucleosome core particle (Wang et al., 2000). The results showed that histone acetylation increases the α-helical content of the histone tails, both in solution and when bound to DNA, suggesting that such an increase is not dependent on the interaction with DNA. Moreover, the α-helical content increased exponentially with the extent of acetylation. It has been shown that acetylation of H4 at K16 regulate the interaction of the H4 NTD with the acidic patch formed by H2A and H2B of neighboring nucleosomes. This interaction is important for mediating condensation of nucleosome arrays (Caterino and Hayes, 2007; Kalashnikova et al., 2013).

Histone H3 is the core histone with the longest NTD and is also coincidentally the one with more PTMs. At least 17 kinds of modification on more than 30 amino acid residues of human histone H3 variants have been reported, including acetylation, ADP-ribosylation, biotinylation, citrullination, crotonylation, formylation, O-GlcNAcylation, glutathionylation, methylation, phosphorylation, propionylation, succinylation, and ubiquitination (Xu et al., 2014).

PTMs act as signals to recruit, evict or repel chromatin regulators. Protein domains that can recognize specific PTMs within histones are known as chromatin readers as they mediate the transduction of the epigenetic signals. The best known methyl-lysine recognition modules include chromo and tudor domains, while the bromodomains recognize acetyl-lysine (Yang et al., 2012).

There are a number of other effector proteins containing chromodomains that recognize unique methyl-lysine marks, and each has a distinct effect on gene expression. Binding of the chromodomains to a methylated lysine residue involves two factors. The first factor consists in the recognition of methylated lysine by an aromatic cage, which appears to be mediated by cation–π interactions between the methylated ammonium group and the side chains of three aromatic residues. The second factor refers to sequence selectivity surrounding the methylated lysine. Apparently, sequence specificity is achieved by β-sheet mediated protein–protein interactions (Hughes et al., 2007).

One well studied modification is trimethylation of H3 at K9, which is associated with epigenetic silencing. This modification is specifically recognized by the chromodomain of HP1. Structural studies revealed that the H3 tail inserts as a β-strand (Figure 3B), completing the β-sandwich architecture of the chromodomain. The methyl-ammonium group is caged by three aromatic side chains, whereas adjacent residues form discerning contacts with one face of the chromodomain (Jacobs and Khorasanizadeh, 2002). Cross-strand interactions between H3 and the chromodomain are thought to determine the sequence specificity of heretochromatin protein 1 (HP1) for the TARKS motif (Eisert et al., 2015). A similar structure has been found in human CBX chromodomains (Kaustov et al., 2011).

Analysis of the crystal structure of the complex of an H3 peptide trimethylated at K4, with the tudor-like domains of Spindlin1 revealed that methyl-lysine recognition involves a hydrophobic cage containing four aromatic residues instead of the typical cage formed by three residues. The surrounding histone sequence is recognized in a distinct manner, involving the amino terminus and a pair of arginine residues of histone H3. It also adopts a β-strand conformation (Figure 3B) (Yang et al., 2012).

A systematic study of the interactions of bromodomains with acetylated histone peptides suggests that multivalent recognition of PTMs is an important and widespread function of chromatin readers. PHD finger is a type of zinc finger domain that can occur as a single finger, but is often found in clusters of two or three accompanied by chromo or bromodomains. Single PHD domains typically recognize up to nine residues of the H3 NTD. A similar binding mode to that of chromodomains, involving the formation of antiparallel beta-sheets between H3 and single PHD domains, has been described for the bromodomain PHD finger 2(BRPF2), BRAF35 HDAC Complex protein (BHC80) and inhibitor of growth family member 5 (ING5) complexes. Recently, structural studies of recombinant constructs of the developmental regulator monocytic leukaemia zinc finger protein (MOZ), containing two tandem PHD fingers (double PDH finger domain, DPF) at the N-terminus followed by a zinc finger (Zn) and HAT domains have brought new insight into the H3 recognition mechanism by PHD domains (Dreveni et al., 2014).

MOZ can acetylate H3K9 and H3K14 but not when H3K4 is trimethylated. Crystallographic analysis of MOZ in complex with H3 peptides has revealed the molecular basis of H3 recognition. Binding to the DPF domain induces α-helical conformation in H3, spanning from K4 to T11 (Figure 3B). The helical structure facilitates sampling of H3K4 methylation status, while the side chain of H3K9 is orientated away from the DPF in a position ideal to allow its modification. Acetylation at K14 further enhanced H3 binding to MOZ, preserving the helical structure. A conformational change mediated through a conserved double glycine hinge (G12/G13) flanking the H3 tail helix allows the acetylated H3K14 side chain, now devoid of charge, to insert into a preformed pocket on the DPF surface (Dreveni et al., 2014).

Interestingly, the tridimensional structure of the DPF domain of BAF45 showed a different mode of recognition of H3K4, involving salt bridges with acidic residues of the DPF domain. Equivalent binding pockets for H3K14ac have been observed, albeit involving different interactions. In this case, no helical conformation has been observed in H3. The alternative conformations of the H3 tail may reflect functional differences between BAF45 and MOZ complexes. The first have no intrinsic enzymatic activity and thus may function uniquely as histone PTM 'reader', while MOZ imposes a structural conformation on the H3 tail to facilitate its acetylation. However, close inspection of available structures have revealed indications of potential helicity in single PHD domains. It has been suggested that the alternative H3 conformations arise from differences in a region, called L1, within the reader module (Dreveni et al., 2014).

1.3. High Mobility Group (HMGs) Proteins

All HMG proteins have similar biochemical and biophysical properties. They are characterized by the presence of a negatively charged intrinsically disordered region at the C-terminal end (Figure 4) (Reeves, 2010). HMGs consist of three different sub-families, HMGA, HMGB and HMGN, which differ from each other in their DNA-binding motifs. They are involved in modulating nucleosome and chromatin structure and orchestrating the efficient participation of other proteins in transcription, replication and DNA repair (Tantos et al., 2012).

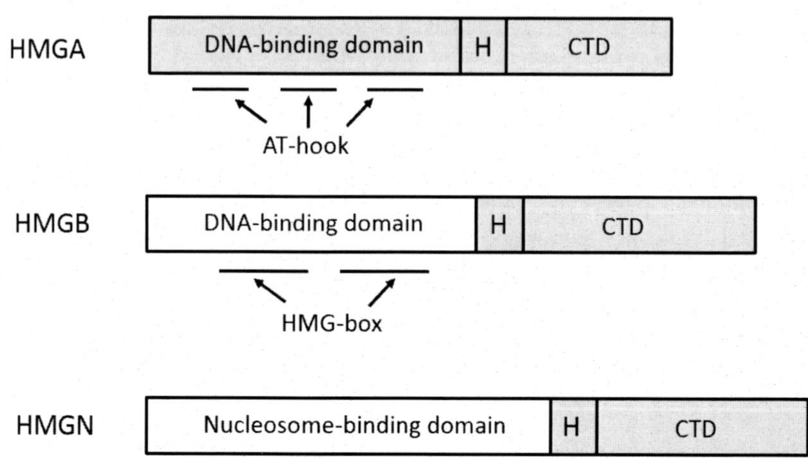

Figure 4. Structural domains of HMGs. All three HMG sub-families share a modular structure with the DNA-binding domain at the N-terminal part of the protein, followed by a hinge region (H) and an intrinsically disordered acidic region at the C-terminal. In light gray, intrinsically disordered regions. Underlined the DNA-binding motifs of HMGA and HMGB.

HMGA proteins have several AT-hook DNA binding motifs, which preferentially bind to the minor groove of short stretches of AT-rich DNA (Figure 4). Most AT-hook domains are formed by a conserved core sequence, Pro-Arg-Gly-Arg-Pro, flanqued by positively charged amino acids (Fonfria-Subiros et al., 2012). In addition to recognizing the structure of the minor groove of AT-rich DNA as a target for binding, HMGA proteins also recognize and bind to non-B-form DNAs with unusual structural features. These binding substrates include bent and supercoiled DNAs and distorted or flexible regions of DNA on isolated nucleosome core particles (Reeves, 2010).

HMGA are completely disordered in their free form and only assume secondary structure upon binding to DNA or other substrates (Huth et al., 1997; Fonfria-Subiros et al., 2012). HMGA is methylated at several arginine residues in cancer cells. The most common site of methylation occurs in the first AT-hook which would probably diminish its affinity for DNA (Fonfria-Subiros et al., 2012).

HMGA proteins can bend, straighten, unwind and supercoil DNA substrates *in vitro* without requiring energy input. These capabilities likely contribute to the role of HMGA in coordinating the formation of a multi-subunit, stereospecific protein–DNA complex called enhanceosome. The assembly of this structure appears to be important for the induction of the expression of immune effector molecules including IFN-β, IL-2 and IL-2α (Reeves, 2010). HMGA proteins are also thought to influence global chromatin structure. They specifically bind to the highly repetitive AT-rich alpha-satellite DNA and to scaffold attachment regions (SARs) (Reeves, 2010). Intrinsic disorder in HMGA may be necessary to allow bending of the DNA and the simultaneous collapse to the lower-energy conformation in the final complex (Dyson, 2012).

Proteins in the HMGB subfamily contain two tandem DNA binding regions called HMG-box domains followed by a 30 amino acid long disordered tail (Figure 4). The HMG-box sharply bends the DNA. It is formed by three α-helices folded together into an L-shaped whose concave surface binds into the minor groove of DNA with limited or no sequence specificity (Hock et al., 2007). It has been suggested that HMGB-induced DNA bending produces an allosteric transition structure that promotes the recognition and binding by other proteins during the formation of functional multiprotein: DNA complexes (Reeves, 2010).

HMGB proteins also bind with high affinity to already distorted DNA structures such as four-way junctions, bulges and kinks. The 3D structure of the entry/exit regions of the nucleosomes has been suggested to resemble a four-way junction. It has also been shown that HMGB proteins bind the DNA segments at the entry/exit of nucleosomes in a similar way to H1. However, in contrast to H1, HMGB facilitates the recruitment of chromatin remodeling proteins that induce nucleosome sliding (Tantos et al., 2012).

HMGB proteins can enhance sequence-specific transcription factor binding to target DNA response elements by a "hit and run" mechanism. In this mode of action HMGB proteins facilitate stable binding of transcription factors to their DNA recognition sites but then dissociates from the ternary complex that is formed, leaving the partner protein stably bound to its DNA substrate (Reeves, 2010). This phenomenon has been described for the interaction of HMGB with several intrinsically disordered transcription factors, including Hox proteins, p53 and steroid hormone receptors (Hill et al., 2012). HMGB1 interacts with the progesterone receptor by an intrinsically disordered region in the hinge domain called C-terminal extension (CTE). It has been proposed that HMGB1 interacts with the CTE of the progesterone receptor (PR) and disrupts a repressive intra-molecular interaction between the CTE and its core DNA-binding domain (DBD), which favors a conformation of the CTE optimal for interaction with the minor groove of the DNA flanking PRE (Progesterone Response Element,). It has been shown that the interaction of HMGB1 with steroid receptors facilitates receptor–DNA binding and recognition of non-consensus HREs (Hormone Response Element) (Hill et al., 2012).

In contrast with the other subfamilies, HMGN proteins are only found in vertebrates. All HMGN proteins contain three important functional regions: a bipartite nuclear localization signal (NLS), a 30 amino acid long nucleosomal binding domain (NBD) and an acidic tail at the CTD, also called regulatory domain (Figure 4). The nucleosomal binding domain (NBD) is well conserved among all HMGN proteins and specifically recognizes the nucleosome core particle both *in vitro* and *in vivo*, a unique feature of this type of proteins (Hock et al., 2007). It contains the invariant octapeptide RRSARLSA, which binds to a negatively charged patch formed by the H2A-H2B dimer. The N-terminal region of the HMGN NBD contacts histone H2B and the DNA approximately 25 base-pairs away from the end of the 147 base pair nucleosomal core DNA, while the C-terminal region of the NBD contacts the DNA near the nucleosomal exit/entry. The C-terminal domain of the HMGN protein contacts the DNA in the two major grooves flanking the nucleosome dyad axis and is in close proximity to the N-terminal tail of H3, which protrudes beyond the periphery of the nucleosomal DNA (Postnikov and Bustin, 2015). The chromatin decondensing activity of HMGN can be, at least in part, attributed to its interaction with the acidic patch of the H2A-H2B dimer and with the N-terminal of H3, since internucleosomal interaction have been shown to be facilitated by the tail of H3 and by interactions between the N-terminal of H4 and the H2A-H2B acidic patch (Postnikov and Bustin, 2015). Unlike the DNA binding domains of the other HMG subfamilies, the tridimensional structure of the NBD of HMGN proteins has yet to be determined (Catez et al., 2010).

Several studies have shown that the levels of expression of HMGN are tightly linked to cellular differentiation, displaying an inverse correlation between the level of HMGN and the differentiation state of cells (Reeves, 2010). HMGN proteins promote the unfolding of higher-order chromatin structure and thereby enhance DNA transcription, replication and repair. As many other IDPs, they are subject to extensive post-translational modifications which influence their chromatin binding properties and their function (Tantos et al., 2012). Acetylation and phosphorylation decrease the interaction of HMGN proteins with nucleosomes. In fact, phosphorylation of HMGN prevents its association with mitotic chromosomes (Reeves, 2010).

1.3.1. The Role of the Intrinsically Disordered C-Terminal Domain of HMGs

A common trait among the three HMG families is the organization of their functional domains (Figure 4). The chromatin-binding domain is at the N-terminal part of the protein, while the acidic region is at the C-terminal separated from the chromatin-binding domain by a hinge region (Figure 4). The C-terminal domain is intrinsically disordered and although is not necessary for chromatin binding, it modulates the interaction, and is likely to be involved in the regulation of the binding *in vivo* (Catez et al., 2010).

All HMGs co-localize with histone H1; HMGA in SARs, HMGB and HMGN at the entry/exit region of the nucleosomes. The overlap between the localization of HMGs and H1 suggests a functionally relevant dynamic competition between these proteins. Experimental evidence indicates that the interaction between HMGs and H1 is mediated by the acidic tail of HMGs and the basic CTD of histone H1, which is also an intrinsically disordered region (Postnikov and Bustin, 2015).

HMGA proteins co-localize with histone H1 at scaffold attachment regions (SAR), which are believed to be cis-acting regulatory elements located at the stem of large loops (domains) of gene-containing DNA. It has been suggested that competition between HMGA1 and histone H1 for binding to AT-rich SAR elements affects chromatin compaction thereby impacting gene transcription (Postnikov and Bustin, 2015).

The acidic tail in HMGB folds over both HMG-boxes, thus masking the DNA binding domains (Catez et al., 2010). Binding to H1 would disrupt the intramolecular interaction of the HMGB tail with the DNA-binding faces of the HMG boxes. A potential consequence of this interaction is enhanced DNA binding by HMGB, with concomitant lower affinity of H1 for DNA. This might facilitate displacement of H1 by HMGB in chromatin, and thus favor transcription (Postnikov and Bustin, 2015).

HMGN is located near the dyad of the nucleosome core particle and promotes chromatin relaxation during cell differentiation. It is possible that negatively charged residues in the C-terminal domain of certain HMGNs, which are positioned near the linker DNA, may interfere with the interactions of H1 at this site (Postnikov and Bustin, 2015).

It seems clear that the dynamic interplay between HMGs and H1 is associated with changes in chromatin structure. Further studies are needed in order to understand how the steady state equilibrium between H1 and HMGs is established and how it changes in response to various internal or extracellular signals.

1.4. Protamines

Protamines are small basic proteins that condense the DNA in mature spermatozoa (Bloch, 1969). Typical protamines, from some fish and mollusks, are of simple composition and very arginine rich, usually in the range 60%-80%. They are characterized by a number of stretches of arginine residues separated by neutral amino acids Gallina from chicken and mammalian protamines are also arginine-rich (Figure 5). The greatest difference between fish and mammalian protamines is the amount of histidine and cysteine. Mammalian protamines have approximately 10% cysteine and up to 20% histidine, whereas fish protamines contain neither of this amino acids. In humans three variants are found. They are highly divergent and one of them, protamine 1, is similar to typical protamines.

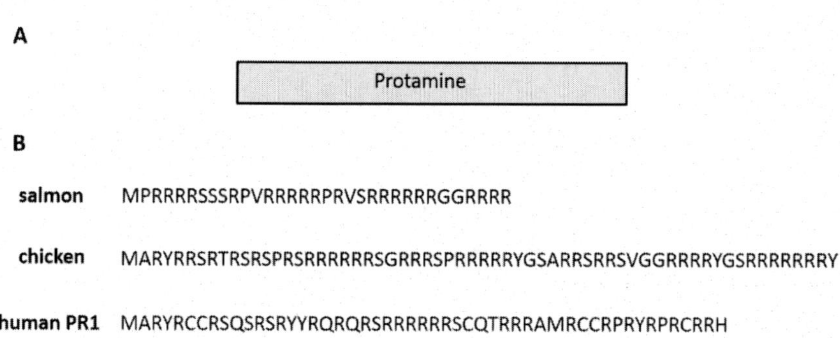

Figure 5. Intrinsic disorder in typical protamines. A. Typical protamines in solution are composed of a single intrinsically disordered domain. B, amino acidic sequence of salmon (Uniprot, P69014), chicken (Uniprot, P15340) and human protamine 1 (PR1; Uniprot, P04553).

Fiber-diffraction diagrams from reconstituted nucleoprotamine and whole sperm heads indicate that DNA molecules are tightly packed in a hexagonal unit cell and that DNA is in a B-like structure. The protein is not visible in these diagrams, likely because of lack of order (Subirana and Puigjaner, 1973). Purified salmine and squid protamine, two typical protamines, had little structure in diluted solution and physiological salt (140 mM NaCl). Addition of TFE revealed the helical trend of both protamines. In 90% TFE, salmine and squid protamine contained ~34% and ~65% α-helix, respectively.

The secondary structure of DNA-bound protamine in cell nuclei was studied by difference FTIR. This technique is particularly well suited to the study of DNA-bound protamine because it is not affected by turbidity. In salmine the α-helix amounted to ~20%, while in squid protamine it reached ~40%. The unordered structure was more abundant in salmine (~40%) than in squid protamine (~20%). Both protamines had ~40% β-turns (Roque et al., 2011).

Protamines are unordered in solution and contain large amounts of defined secondary structure when bound to DNA. They thus belong to IDPs with coupled binding and folding. The remaining unordered structure may help to accommodate the protamine molecules in the space left by tightly packed DNA molecules.

1.5. Methyl-CpG-Binding Protein 2 (MeCP2)

MeCP2 is a chromatin architectural protein, highly abundant in neuronal cells, with key roles in post-natal brain development in humans. Missense, nonsense and frame-shift mutations in MeCP2 sequence cause the Rett Syndrome (RTT). MeCP2 is composed of six biochemically distinct domains (Adams et al., 2007): the N-terminal domain (NTD; residues 1–78), the methyl-DNA binding domain (MBD; residues 79–167), the intervening domain (ID; residues 168–205), the transcription repression domain (TRD; residues 206–309), the C-terminal domain α (CTD; residues 310-355) and the C-terminal domain β (CTDβ; residues 356–486) (Figure 6A). Pathogenic RTT mutations are found in each MeCP2 domain suggesting that all MeCP2 domains in some manner contribute to the function of the full-length protein (Hite et al., 2012).

MeCP2 can preferentially recognize methylated DNA, by its MDB, and act as a methylation-dependent transcriptional repressor. However, MDB and other domains (ID and CTDα) bind to unmethylated DNA and MeCP2 is also associated with actively transcribed genes (Hite et al., 2009). The experimental evidence suggests a model in which MeCP2 functions as an architectural chromatin protein with both a positive and a negative effect on transcription.

To date, structural information is only available for the MBD (amino acids 79–167) alone and in complex with a short fragment of methylated DNA (Wakefield et al. 1999; Ho et al., 2008). The structured core consists in a wedge made up of a 3 stranded anti-parallel β-sheet on one side, with one α-helix on the C-terminal side (Hite et al., 2009). Biophysical and protease digestion experiments have established that native MeCP2 is an intrinsically disordered protein. Circular dichroism (CD) of recombinant human MeCP2 showed that full-length protein was approximately ~35% β-strand/turn, 5% α-helix, and almost 60% unstructured.

A

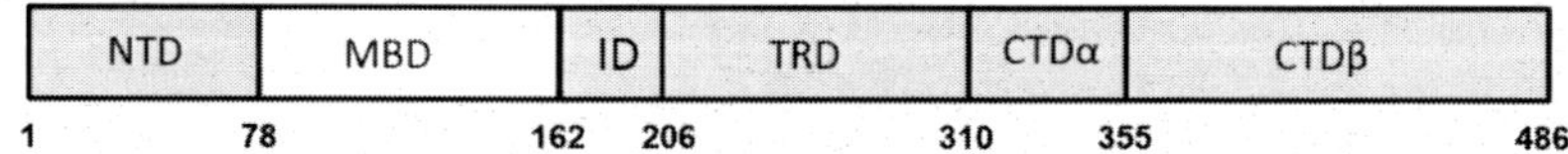

B

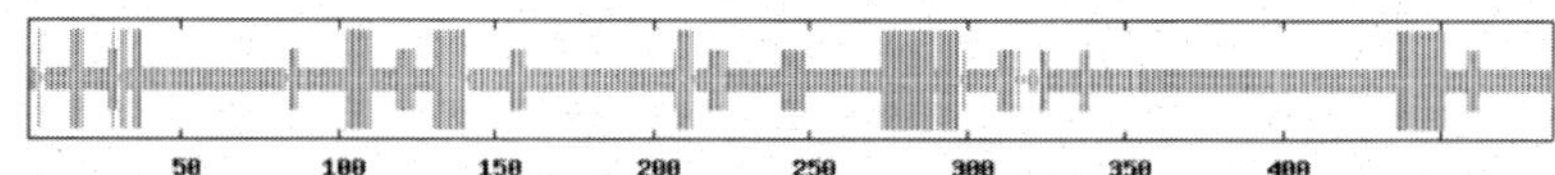

Figure 6. Intrinsic disorder in human MeCP2. A. Structural domains of human MeCP2 (Uniprot, P51608). Human MeCP2 has six domains: the N-terminal domain (NTD), the methyl-binding domain, the intervening domain (ID), the transcription repression domain (TRD), the C-terminal domain α (CTDα) and the C-terminal domain β (CTDβ). In light gray, intrinsically disordered regions. B. Consensus secondary structure prediction for MeCP2. The short lines represent unstructured residues, the lines of middle length lines regions in β-structure and the long lines regions in α-helix.

MeCP2 is predicted to have short stretches of order interspersed between long stretches of internal disorder over the length of the entire polypeptide chain. Prediction analysis suggests that some regions of MeCP2 have propensity to fold in α-helix (Figure 6B). CD analysis indicates that approximately two-thirds of the unstructured residues present in full-length MeCP2 were converted to α-helix in 70% TFE. In these conditions, the unstructured residues of the N-terminal (NTD) and C-terminal (CTD) domains were partially converted to α-helix, while the central transcription regulation domain (TRD) became almost completely α-helical. This indicates that TFE induced extensive coil-to-helix transitions in the full-length protein (Hite et al., 2012).

From a functional perspective, TFE/water solutions may mimic the environment found at the binding interfaces between IDPs and other macromolecules. IDPs often gain secondary structure concomitant with macromolecular interactions. In the case of MeCP2, several interaction partners have been identified; the NTD interacts with the protein HP1, the TRD with many different co-repressor proteins and unmethylated DNA, and the CTD with unmethylated DNA, chromatin, and RNA splicing proteins.

Moreover, some proteins, like CDK5L, interact with multiple MeCP2 domains (Hite et al., 2012).

CD measurements showed that about 40% of MeCP2 was structured. This data usually is understood as if the calculated value is present in the protein in the form of stable three-dimensional structure. However, in the case of MeCP2 mass spectrometry-based hydrogen/deuterium exchange (H/DX) experiments have shown great conformational flexibility throughout the protein, even the MBD. Binding to methylated and unmethylated DNA slowed the exchange rate only of the MBD. These results indicate that the full-length MeCP2 rapidly samples many different secondary structures and equilibrates between multiple tertiary structures, even when bound to DNA. The functional advantages of the extreme conformational flexibility and fast conformational sampling exhibited by MeCP2 could provide biochemical basis for establishing protein-protein interactions with multiple partners and it could also facilitate the formation of conformationally malleable higher-order macromolecular complexes (Hansen et al., 2011).

Studies in chromatin have shown that MeCP2 forms a well-defined complex with nucleosomes, without displacing the core histones. The results showed that MeCP2-induced conformational changes in the nucleosome, similar to those induced by histone H1 and that it also promoted local compaction. In fact it appears that MeCP2 competes with histone H1 for binding to linker DNA, and is able to displace H1 (~40% of H1), mainly from methylated regions (Thambirajah et al., 2012).

MeCP2 protects ~11 bp of extra-nucleosomal DNA as seen from micrococcal nuclease digestions, suggesting that it binds near the nucleosomal dyad axis. The molecular dimensions obtained from SAXS experiments demonstrate that MeCP2, which is extended and disordered in its free state, assumes a less extended conformation upon interaction with nucleosomes containing linker DNA, but remains extended when interacting with nucleosomes without linker DNA. This suggests that additional DNA binding domains in MeCP2 become engaged upon the interaction with linker DNA and nucleosomes (Yang et al., 2011).

2. TRANSCRIPTION FACTORS

One of the main functions associated with intrinsic disorder is transcription regulation. The use of the predictor of natural disorder regions (PONDR) revealed that more than 82% of the transcription factors contain intrinsically disordered regions. Analysis of the distribution of disorder within the transcription factor datasets showed the degree of disorder in transcription activation regions is much higher than that in DNA-binding domains (Liu et al., 2006).

Eukaryotic transcription factors recognize specific DNA sequences and recruit the transcription machinery. Usually, they consist of several structural domains including a DNA-binding domain (DBD) and a transcriptional activation domain, also called transactivation domain (TAD). Both DBDs and TADs have high level of disorder (Tantos et al., 2012). Zinc-fingers, basic helix-loop-helix (bHLH), basic leucine zipper (bZIP), and AT-hook domains are the most representative DNA-binding motifs present in transcription factors.

Zinc-fingers are small protein domains (~30 residues) that form finger-like protrusions that make contacts with the target molecule. They can be classified into several classes according to the type and order of the residues coordinating with the zinc or on the overall shape of the protein backbone in the folded domain (Krishna et al., 2003). Zinc finger domains are often found in clusters, where individual fingers can have different binding specificities.

Zinc fingers belong to the most highly ordered DNA binding protein motifs (Tantos et al., 2012). However, initial studies with the zinc finger motif of transcription factor IIIA of *Xenopus laevis* showed that the motif was mostly disordered in the absence of zinc but became stably folded in the presence of this ion (Frankel at al., 1987). Although zinc fingers rarely undergo conformational changes upon binding to their targets in nuclear hormone receptors, a short region between the two helical segments experiments a disorder to order transition upon binding to DNA.

bZIP transcription factors are the second largest family of eukaryotic transcription factors. It includes well studied TFs such as CREB (cAMP response element-binding protein), c-fos and c-jun. bZIP domains have two distinct regions, the N-terminal basic region and the C-terminal leucine zipper region. They bind to DNA as dimers. The basic region of each monomer makes specific contacts with cognate DNA half-sites, while the leucine zipper regions promote dimerization. The crystal structure of the dimeric bZIP in complex with DNA is formed by continuous α-helices in each monomer (Figure 7) (Schumacher et al., 2000; Das et al., 2012).

The basic regions of bZIP comprise ~28 residues and prediction studies suggest that these regions are almost completely disordered. Like AT-hooks, the basic region of the bZIP domains undergoes coupled binding and folding (Tantos et al., 2012). They are mostly unstructured in solution but acquire secondary structure in form of α-helix upon DNA-binding (Figure 7). In fact, DNA binding and coupled folding of the basic regions can occur irrespective of the presence of the leucine zipper region (Hollenbeck et al., 2002). Atomistic simulations and circular dichroism measurements have shown that the monomeric basic regions have α-helical properties that vary among the different basic regions. This behavior is sequence specific and appears to be determined by the 8-residue N-terminal fragment which acts as a modulator for the conformational ensembles of the whole region (Das et al., 2012).

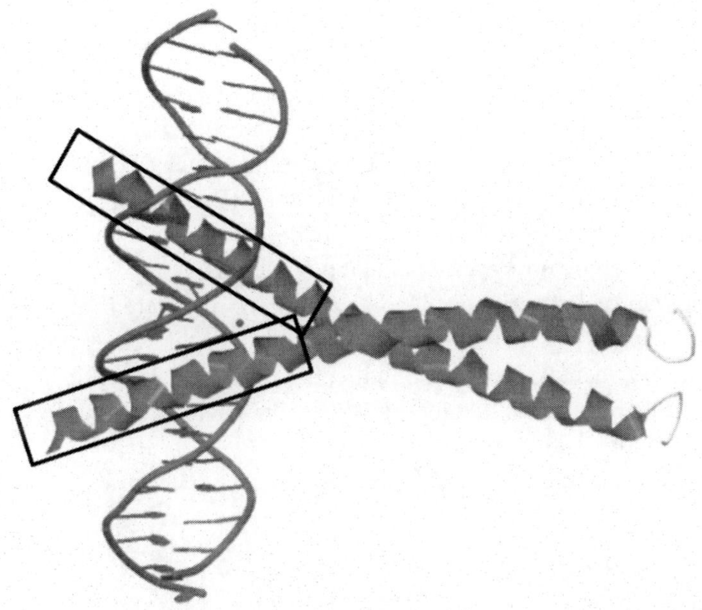

Figure 7. Crystal structure of the bZIP domain of the cAMP response element-binding protein (CREB) in complex with the cAMP response element (CRE) DNA (PDB: 1DH3, Schumacher et al., 2000). The black square corresponds to the basic regions, which are intrinsically disordered and fold upon interaction with DNA.

Some transcription factors have more than one DNA-binding domain that often displays different affinities for DNA (Vuzman and Levy, 2012). In occasions one domain binds to DNA in a sequence specific manner, while the other has nonspecific interactions with DNA, as it happens in p53. The domains with nonspecific binding are highly positively charged, with much smaller separation between the charged clusters than in globular proteins. They are also referred as disordered tails. Their function in TFs has been associated with optimal DNA search and increase of the capture radius of the molecule. In multidomain TFs, two DNA-binding domains can be tethered by a flexible linker, which can increase their affinity for DNA, resulting in higher propensity for sliding along the DNA. Post-translational modifications in DNA-binding domains can also modulate the affinity constant of the interaction (Vuzman et al., 2012).

Transcriptional activation domains (TAD) are highly rich in disorder promoting amino acids, but also contain small hydrophobic motifs which appear to be critical for their activity (Dyson, 2012). They mediate protein-protein interactions and usually binding is accompanied by some gain of structure, suggestive of a disorder to order transition. This coupled binding and folding can be achieved by different mechanisms: induced fit, when folding of the IDP occurs after association with the target, or conformational selection, when the binding involves pre-formed elements of secondary structure (Arai et al., 2015). Sometimes parts of the IDP remain unstructured after binding their targets forming the so called fuzzy complexes (Wright and Dyson, 2015). Structural flexibility of the TAD enables binding to multiple targets and can be further modulated by post-translational modifications.

The group of transcription factors containing disordered regions is quite large and includes proteins involved in the control of the cell cycle, regulators of the developmental programs and cell signaling. As an example, the functional role of intrinsic disorder in p53 and in nuclear hormone receptors is described in more detail.

2.1. p53

Since the discovery of p53 more than 30 years ago, multiple functions for this protein have been revealed. It is a well-known tumor suppressor, which is frequently mutated in human cancer. It is also a transcription factor capable of inducing cell-cycle arrest and apoptosis. Recently reported new roles for p53 include the ability to regulate metabolism, fecundity, and various aspects of differentiation and development (Levine and Oren, 2009; Vousden and Prives, 2009).

Human p53 has 393 residues and contains several distinct domains (Figure 1). The N-terminus (1-67) corresponds to the transactivation domain (TAD), followed by a proline rich domain (68-98). The transactivation domain interacts with Mdm2 and components of the transcription initiation complex (Vousden and Prives, 2009; Weinberg et al., 2004a). Residues 98-303 are part of the core DNA-binding domain, which is responsible of the DNA-sequence specificity of p53 binding. Next, the linker region, which comprises residues 304-324 and connects the core region with the tetramerization (Tet) domain (325-355). The formation of a tetrameric complex is associated with the bending of the p53 DNA recognition element and also plays a role in modulating p53 affinity to its target sequences and in stabilizing the nucleoprotein complex (Weinberg et al., 2004b). The C-terminal domain (363-393) is a lysine-rich domain, which mediates non-specific DNA-binding and modulates p53-DNA-binding properties (Figure 8A) (Vousden and Prives, 2009; Weinberg et al., 2004a).

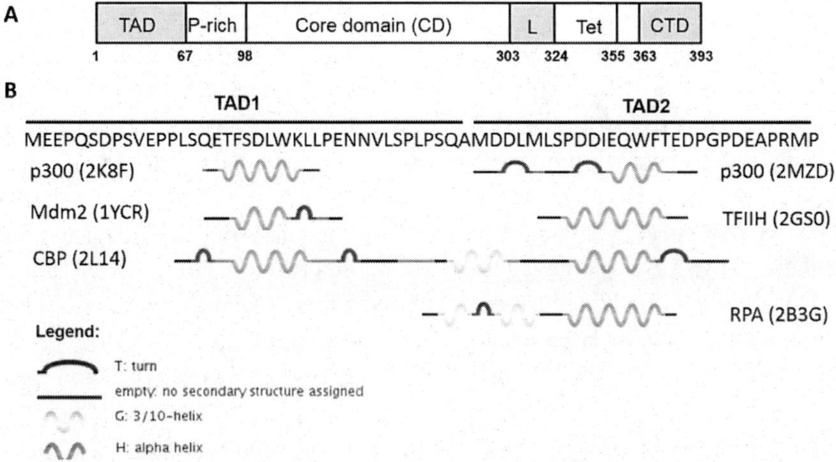

Figure 8. Structural domains and folding of p53. A. Domains of p53. Transcriptional activation domain or transactivation domain (TAD), proline-rich (P-rich) domain, core domain (CD), linker region (L), tetramerization domain (Tet) and C-terminal domain (CTD). In light gray, intrinsically disordered regions. B. Secondary structure motifs in the transactivation domain of p53 in complex with different proteins. The sequence corresponds to the residues from 1 to 67 of human p53 (Uniprot, P04637). The upper horizontal bar denotes the transcriptional subdomains TAD1 and TAD2. In parenthesis the four digit code corresponding to the PBD entry code of the tridimensional structure.

It is difficult to crystallize full-length p53 or to determine its structure using NMR because 40% of the protein is intrinsically disordered. Specifically, three of the above mentioned regions are intrinsically disordered, the N-terminal domain, the linker region and the C-terminal domain.

2.1.1. IDRs of p53 and Protein-Protein Interactions

2.1.1.1. Transactivation Domain (TAD)

The N-terminal domain of p53 contains the transcriptional activation domain or transactivation domain. This region is intrinsically disordered, highly acidic and can be subdivided in two subdomains, TAD1 (1-39) and TAD2 (40-67) (Teufel et al., 2007) (Figure 8A). Although the N-terminal domain of p53 is mostly unstructured in solution, it has a weak propensity to form transient helical structure between residues 18 and 26 and two turns (residues 40–44 and 48–53), as revealed by NMR (Lee et al., 2000). The TADs have been shown to interact functionally with a variety of transcriptional regulatory proteins like Mdm2, CBP, RPA70 and the transcription factor II H (TFIIH) (Di Lello et al., 2006) (Figure 8B).

In normal cells, Mdm2 and p53 form a negative feedback loop that helps to limit the growth-suppressing activity of p53. Its catalytic activity, E3-ubiquitin-ligase, results in p53 ubiquitination and subsequent degradation. Mdm2 is also capable to inhibit p53 activity by concealing its transactivation domain. X-ray results showed that Mdm2 interacted with a stable folded amphipathic helix (residues 18 to 26), with the three hydrophobic amino acids, F19, W23 and L26, inserted deep within the Mdm2 cleft (Kussie et al., 1996) (Figure 8B). Later studies showed that the second nascent turn (49-54) also interacted with Mdm2 but with weaker affinity than the α-helix (Chi et al., 2005). NMR analysis revealed that upon Mdm2 binding the turn II motif becomes more ordered from a nascent turn to a well-defined one consisting of a complete turn of an amphipathic helix. One face of this one-turn helix contains three hydrophobic residues, I, W and F, which favor the binding to the hydrophobic helix binding pocket in Mdm2. Interestingly, the helix and turn II affect different residues of the same helix binding pocket of Mdm2.

Other relevant partners of p53 are the closely related coactivators p300 and cAMP response element binding protein (CBP). These proteins have acetyltransferase activity, which is responsible for histone acetylation at the p53 targeted promoters and thereby mediate local chromatin unwinding. They also acetylate six lysine residues at the C-terminal domain of p53 which further activate and stabilize the protein (Feng et al., 2009).

The N-terminal domain of p53 has been shown to interact with several binding domains of CBP/p300 including Taz1, KIX, Taz2 and the C-terminal nuclear receptor coactivator binding domain (NCBD) (Feng et al., 2009). Binding to CBP is dominated by interactions with TAD2, while binding to Mdm2 is mediated primarily by interactions involving the TAD1. The p53

TAD can bind simultaneously to Mdm2 and to any one of CBP domains, through the TAD1 and TAD2 motifs, respectively, to form a ternary complex (Lee et al., 2010a).

The transcriptional adaptor zinc-binding domains (Taz1 and Taz2) of CBP/p300 are also capable of binding both TAD subdomains of p53 (Jenkins et al., 2009; Polley et al., 2008; Teufel et al., 2007). The structure of the complex between p300-Taz2 and p53-TAD1 has been determined by NMR (Feng et al., 2009). In the complex, the p300 Taz2 domain forms four core α-helices with three HCCC-type zinc-binding motifs, whereas only residues 15-27 from TAD1 are stably folded in α-helix (Figure 8B). The helical region of p53 interacts with helices α1, α2, and α3 of Taz2. The p53 residues interacting with p300 are primarily hydrophobic, F19, L22 and L25, and are located within the helical region. Contributions from charged and polar residues located throughout TAD1 were also observed.

The KIX domain was first characterized as the region of CBP responsible for binding the phosphorylated kinase inducible activation domain (pKID) of CREB. KIX has two distinct binding surfaces suggesting that KIX is capable of binding two transcriptional activation domains simultaneously to form a ternary complex (Lee et al., 2009). Mapping the interactions of the p53 TAD with KIX has shown that the p53 TAD simultaneously occupies the two distinct binding sites that have been identified on the KIX domain and efficiently competes for these sites with other known KIX-binding transcription factors (Lee et al., 2009). The structure of CBP KIX domain in complex with p53-TAD has not been determined. However, the available structures of the KIX domain in complex with other members of the family of the 9 amino acid transactivation domains, MLL and HEB, have allowed generating a structural model for p53-TAD/CBP-KIX interaction. In this model both TAD subdomains would form helical elements and the hydrophobic doublets present in each subdomain, L22/W23 in TAD1 and W53/F54 in TAD2, would be critical in the interaction with the hydrophobic grooves of the KIX domain (Piskacek et al., 2015).

TAD of p53 is an intrinsically disordered region and the unbound NCBD is not entirely unstructured but forms a helical state with features of a molten globule. NMR experiments indicate that upon binding to each other, both NCBD and TAD undergo a disorder to order transition (Yu at al., 2013). The complex structure has 5 α-helices, three belong to NCBD and the other two belong to TAD1 (residues 19-25) and TAD2 (residues 47-53) subdomains of p53, respectively (Figure 8B). Complex formation is driven largely by hydrophobic contacts that form a stable intermolecular hydrophobic core, involving the 2 Trp and 2 Phe residues within both α-helical elements. A salt bridge between D49 of p53 and R2105 of NCBD may contribute to the binding specificity (Lee et al., 2010b). Molecular dynamics simulations, high temperature unbinding kinetics and room temperature landscape analysis suggest that the molecular recognition of p53 and NCBD may involve a local induced fit and global conformational selection mechanisms (Yu et al., 2013).

The interaction of p53 and TFIIH is associated with the ability of p53 to activate transcriptional elongation. TFIIH is a general transcription factor composed of ten subunits. One of these subunits, p62 and its yeast counterpart Tfb1, has been shown to interact with the transactivation domain of p53 and thus mediate its recruitment to the TFIIH complex. *In vitro* binding results indicated that p62/Tfb1 interacted specifically with TAD2. Phosphorylation of two residues within TAD2, S46 and T55, led to an enhancement of p53 binding to p62/Tfb1. The NMR structure of the complex Tfb1-p53 showed that residues from 47 to 55 of TAD2 formed an amphipathic α-helix, which interacts with a surface of Tfb1 formed by β strands β5, β6, and β7 and the loop connecting β5 to β6 (Di Lello et al., 2006) (Figure 8B).

Upon DNA damage, p53 is activated and induces a cellular response that includes the transcriptional activation of genes involved in cell cycle arrest and apoptosis. p53 is also part of a network of protein-protein and protein-DNA interactions that regulate DNA repair. One of the p53 partners is the replication protein A (RPA). The interaction of p53 and RPA mediates suppression of homologous recombination (Bochkareva el at., 2005). It is also linked with DNA repair and disruption of p53 and RPA complexes after DNA damage. It is thought to coordinate DNA repair with the p53-dependent checkpoint control. The 1.6-Å crystal structure of a chimeric protein containing the N-terminal 120 residues of RPA70, followed by p53 residues 33–60, showed that the p53 TAD2 folded into two helices, H1 (residues 41–44) and H2 (residues 47–55) upon binding (Figure 8B). H2, an acidic amphipathic helix, appears to be the major determinant of the interaction and binds in the deep basic cleft corresponding to the nucleic acid-binding pocket of the oligonucleotide oligosaccharide-binding (OB) fold. The orientation of this helix is stabilized by electrostatic and hydrophobic interactions. Additionally, p53-AP interaction is affected by DNA damage signals such as ssDNA and hyperphosphorylated RPA32 (Bochkareva el at., 2005). The RPA-binding region of p53 contains two functionally important phosphorylation sites, phosphorylation at S37 is involved in DNA damage response and phosphorylation at S46 is associated with p53-mediated apoptosis. Recent studies showed that dissociation of the complex p53-RPA required the synergistic effect of hyperphosphorylation of RPA32 and p53 phosphorylation at S37 and at S46 (Serrano et al., 2013).

It is commonly known that p53 is the most frequently mutated protein in human cancers. While 80% of p53 cancer mutations are located in the core DNA-binding domain, some mutations have been found in the TAD region. Recent simulations suggest that some cancer-associated mutations affect the local and distal helical propensities of p53-TAD (Ganguly and Chen, 2015). Mutations K24N and W53G appear to decrease TAD1 and TAD2 helical propensities, respectively. W53G also leads to a slight decrease in helicity within TAD1 subdomain. The higher helicity changes in regions sequentially distal from the mutation sites were observed in N29K/N30D. These mutations reduce the helical propensities of TAD2 by ~50%. However, little biophysical data is available on how TAD cancer mutants may perturb the interaction of p53 with various regulatory proteins. The only molecular data available is that the mutation K24N does not significantly affect Mdm2 binding (Zhan et al., 2013), but its impact on binding to CBP domains or other ligands is not known.

2.1.1.2. C-Terminal Domain (CTD) of p53

The CTD of p53 is involved in both protein-DNA and protein-protein interactions. Folding of the p53 CTD upon DNA interaction have not been studied in detail. The CTD contains a large number of residues that show chemical shift perturbations by NMR upon binding to DNA, suggesting that the binding may be coupled to folding. Moreover, the changes in the chemical shifts are located among residues 365–382, a region that binds S100B($\beta\beta$) and the bromodomain of CBP. This region has been shown to fold after binding to regulatory proteins (Weinberg et al., 2004a).

In addition to DNA binding, the CTD of p53 interacts with Sir2, S100B($\beta\beta$), CBP, SMYD2, the DNA repair factor 53BP1 and many other proteins (Hoff et al., 2006; Mujtaba et al., 2004; Rustandi et al., 2000; Roy et al., 2010; Wang et al., 2011). The tridimensional structures of p53 CTD peptides in complex with these regulatory proteins have been obtained by NMR or X-ray crystallography and show that the same region of the p53 CTD is able to fold in different conformations (Figure 9), indicating its structural flexibility.

The crystal structure of the complex of *Thermatoga maritima* Sir2 bound to an acetylated p53 peptide (372-389 acetylated at K382) and NAD$^+$ has shown that Sir2 binds the p53 peptide by forming backbone hydrogen bonds with loops flanking the peptide binding cleft, giving rise to a three-stranded enzyme-substrate β-sheet, with the peptide as the central strand. This β-sheet structure orients the acetyl lysine of the peptide substrate in a large hydrophobic cleft (Hoff et al., 2006). The mammalian orthologue, SIRT1, is able to deacetylate p53 *in vivo* suggesting a role in tumor suppression (Lee and Gu, 2013).

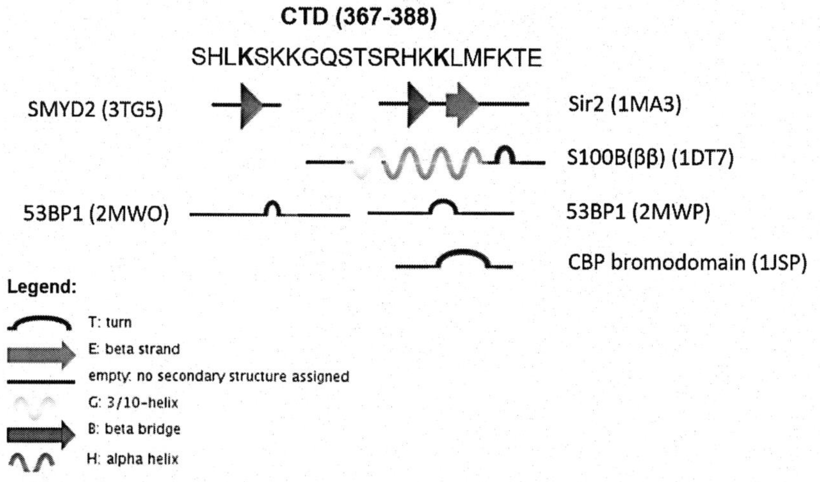

Figure 9. Secondary structure motifs found in the C-terminal domain (CTD) of p53 in complex with different proteins. The sequence corresponds to the residues from 367 to 388 of human p53 (Uniprot, P04637). In parenthesis the four digit code corresponding to the PBD entry code of the tridimensional structure. Lysines in bold are post-translationally modified in some of the structures: K382 is acetylated in 1MA3 and 1JSP; K370 and K382 are dimethylated in 2MWP.

p53 CTD contains several posttranslational modification (PTM) sites that are important for regulation of p53 activity. Calcium-dependent binding of dimeric S100B(ββ) to p53-CTD blocks access to these PTM sites and disrupts the p53 tetramer inhibiting p53 activation. NMR studies of the complex of S100B(ββ) and a p53 C-terminal peptide (367-388) have shown that although this region is unstructured in its native form, residues 376-387 become helical upon binding S100B(ββ) (Figure 9). The residues of p53 in direct interaction with S100B are F385 involved in hydrophobic interactions and two basic residues, R379 and K386, which form a salt bridge with two glutamic residues of S100B(ββ) (Rustandi et al., 2000). Closer examination of the NMR data suggested that the complex might not be as stable as previously thought (McDowell at al., 2013). Extensive atomistic simulations revealed that p53CTD-S100B(ββ) might be instead a fuzzy complex where many conformational sub-states coexist in the bound state. Specifically, helix unwinding at the p53-CTD allows key hydrophobic residues (L383 and F385) to make more extensive contacts with S100B($\beta\beta$). The N-terminal segment, although highly helical, displays great flexibility in its packing with S100B($\beta\beta$).

In response to DNA damage, p53 is acetylated in multiple residues of the CTD by CBP/p300. *In vivo* studies show that lysine acetylation of p53 does not result in the enhancement of its DNA binding ability but, rather, promotes the recruitment of coactivators, leading to histone acetylation and transcriptional activation of target genes (Barlev et al., 2001). The bromodomain of CBP binds specifically K382ac and is responsible for p53 acetylation-dependent coactivator recruitment after DNA damage. NMR analysis of the three-dimensional structure of the CBP bromodomain in complex with a K382ac-p53 peptide (residues 367–386) showed that the structure of the peptide (residues 381–385) was well-defined and consisted of a β turn-like conformation with the acetylated K382 located at the beginning of the turn (Figure 9) (Mujtaba et al., 2004).

High-resolution crystal structure of the full-length SMYD2, a lysine methyl transferase, in ternary complex with S-adenosylhomocysteine and the p53 peptide (residues 368–375) revealed that the p53 peptide binds to a deep pocket of the interface between the catalytic SET domain (1–282) and the p53 C-terminal domain with an unprecedented U-shaped conformation, with the residues (aa 370–372) located at the bottom of the U-shaped conformation in contact with the enzyme core (Figure 9) (Wang et al., 2011).

It has been found that the p53K382me2 levels increase upon DNA damage, and that the damage triggers an acute mobilization of the DNA repair factor p53 binding protein 1 (53BP1) to the double-strand break sites, which in turn can promote accumulation of p53. The dimethyl-lysine recognition at positions 370 and 382 of p53 by the Tudor domain of 53BP1 involves the retention of the modified lysine in the hydrophobic pocket, which induces turn-like conformations in the CTD with the modified residues near its center (Figure 9) (Roy et al., 2010).

2.1.2. Role of the Disordered Domains in p53 Conformation and Target Recognition

The linker region is the shortest IDR of p53. This disordered region mediates the cross-talk between the core domain and the Tet domain in absence of a physical interphase between them. p53 is a tetrameric transcription factor, formed by a dimer of dimers where the Tet domain forms a symmetrical tetramer made of two tight dimers stabilized by an anti-parallel β-sheet and helix–helix interactions. Four core domains (one from each p53 monomer) interact with cognate DNA, comprising two decameric half-sites spaced 0-20 bp apart. The flexibility of the linker region between the core domain and the Tet domain allows for specific recognition of the target sites with different spacing (Khazanov and Levy, 2011).

Different conformations of the tetramer when bound to DNA have been experimentally determined and also predicted by computational models (Khazanov and Levy, 2011). Atomistic models suggest that the axis of the Tet domain can run parallel or perpendicular to the DNA molecule depending on the salt concentration, crowding of the environment or the presence of one or two DNA molecules. The disordered linker also allowed p53 core and Tet domains to adopt different conformations when bound to DNA.

The functional role of the p53 C-terminal domain has been a subject of great controversy. Initial results suggested that it negatively regulated the DNA-binding of the core domain to its specific site on the DNA. This conclusion was based on the fact that deletion of the CTD, phosphorylation of a serine residue or the use of an antibody against the CTD caused an increase in the binding of the core domain to DNA *in vitro* (Hupp et al., 1992). These results led to the hypothesis that the nonspecific interaction of the C-terminal domain with the DNA interferes with the ability of the core domain to bind to its specific target sequence until relevant signals cause modifications in the C-terminal domain that alleviate its negative regulation of core DNA binding (Tafvizi et al., 2011).

In contrast, *in vivo* studies showed that p53 with a truncated CTD was considerably less efficient at binding and transactivating targets. The C-terminal domain was also required for efficient target recognition using long circular DNA and for the one-dimensional translocation along the DNA molecule. Therefore, these results suggested that the C-terminal domain had a positive regulatory role in DNA-binding (McKinney et al., 2004).

Recently, single-molecule imaging tools have been used to study the role of the C-terminal domain in p53 target recognition (Tafvizi et al., 2011). The results showed that while the core domain of p53 translocates on DNA via a hopping mechanism, the C-terminal domain translocates on DNA via a sliding mechanism. These results suggest that p53 moves on DNA through a two-state mechanism of Search and Recognition. The two-state mechanism suggests that both fast search and sequence-specific recognition can be achieved if the protein has two distinct conformational states: a Search state characterized by largely nonspecific binding and fast sliding, and a Recognition state in which a protein binds to DNA in a sequence-specific manner, while unable to slide. In the case of p53 each role would be accomplished by a different domain. This model proposes that the Search state corresponds to a conformation where the C-terminal domain is bound to DNA while the core domain is not, allowing for unspecific binding and fast translocation. In the Recognition state, the core domain binds to DNA, allowing for specific recognition. Conformational switching between the two states allows both sufficiently fast translocation and specific DNA binding. This model reconciles contradictory results about the role of p53 CTD. Therefore, the C-terminal domain would have a negative thermodynamic effect upon specific binding by sequestering p53 to unspecific sequences, while it would have a positive kinetic effect at facilitating the search of the target sequence. The disordered and charged C-termini of p53 would accelerate search speed. Additionally, the large capture radius of the C-tail allows interactions with distant DNA molecules, which may increase the localization of p53 to DNA.

2.1.3. Regulation of p53 by PTMs

The p53 protein is controlled by many different forms of post-translational modifications, including ubiquitination, phosphorylation, acetylation, methylation, sumoylation, and neddylation (Reed and Quelle, 2015). These modifications can dictate the p53 response to diverse cellular signals and help determine its physiological activities. Similar to other IDPs, in p53 the PTMs are clustered mainly within intrinsically disordered regions and modulate its function. However, if these PTMs also modulate p53 structure is not known.

Of particular interest is how multisite phosphorylation functions as a rheostat to enhance binding to CBP domains in a graded manner. Phosphorylation of p53 at T18 following DNA damage impairs binding to the N-terminal region of Mdm2, while enhancing binding to the CBP TAZ1, TAZ2, and KIX domains. Multisite phosphorylation at S15, S20, S33, S37 and S46 enhances binding of p53 to CBP/p300 through a graded response that is proportional to the number of phosphoryl groups. Multiple phosphorylation enhances the ability of p53 to compete with cellular transcription factors for binding to limiting amounts of CBP/p300, and may control the nature and extent of the p53 response to prolonged genotoxic stress (Lee et al., 2010b).

In summary, intrinsic disorder is essential for p53 function, although in each region it has a different role. In the N-terminal domain intrinsic disorder is associated with protein-protein interactions and allows for multiple target recognition, which often involves disorder to order transitions. In the linker region its function arises from the disordered state, which allows for structural diversity in DNA sequence recognition and binding. The linker region also mediates the cross-talk between the core and Tet domains without any physical interphase. In the C-terminal domain intrinsic disorder is important in protein-protein interactions that involve the folding of the same region of p53 in quite different secondary structures. An intrinsically disordered CTD also appears to be important in the p53 mechanism for searching its specific DNA sequence in the genome.

2.2. Nuclear Hormone Receptors (NHRs)

Nuclear hormone receptors are the largest family of metazoan transcriptional regulators. They share an architecture that consists in a variable amino-terminal domain (NTD), a highly conserved zinc-finger DNA-binding domain (DBD), a connecting hinge region and a less conserved ligand-binding domain (LBD) (Figure 10) (Khorasanizadeh and Rastinejad, 2001). On the basis of their active conformation the NHRs can be classified in three groups: the receptors that bind DNA as homodimers, heterodimers or monomers. The first group includes receptors for estradiol (ER), progesterone (PR), androgens (ARs), glucocorticoids (GRs) and mineralocorticoids (MRs). The second major group includes receptors for all-trans retinoic acid (RAR), vitamin D3 (VDR) and thyroid hormone (TR), while members of the third group include receptors such as NGFI-B, RevErb, ERR2, ROR and SF-1. Some members of the family remain 'orphans', with no specific ligands yet identified (Khorasanizadeh and Rastinejad, 2001).

The classic mechanism of steroid action, based on the domain model, assumes that the ligand bound receptor enters the nucleus where it binds to its response element DNA site. This allows the formation of a transcription initiation complex either directly or indirectly through subsequent binding to coregulatory proteins. To a large extent, the composition of the receptor: coregulator complex assembly determines the final outcome of the target gene regulation by the receptor (Kumar and Litwack, 2009).

There appears to be an intimate relationship between NHR function and structural disorder. Bioinformatic analysis revealed that NTD and the hinge region had the highest levels of predicted disorder (Krasowski et al., 2008).

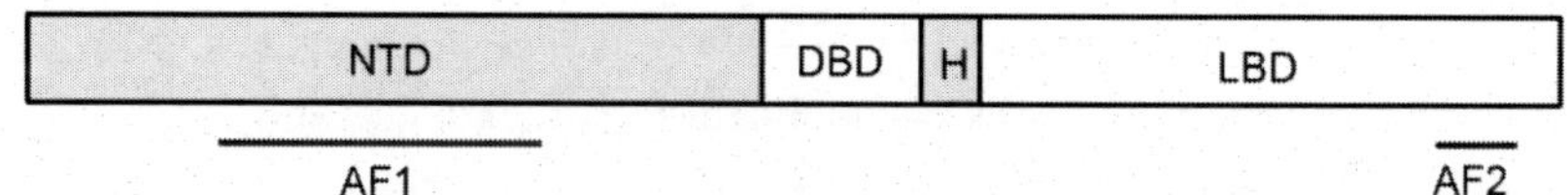

Figure 10. Structural domains of nuclear hormone receptors (NHRs). NHRs contain four domains; the N-terminal domain (NTD), the DNA-binding domain (DBD), the hinge domain (H) and the ligand-binding domain (LBD). In light gray, intrinsically disordered regions. Underlined within the NTD and the LBD the two major transactivation domains AF1 and AF2, respectively.

2.2.1. Intrinsic Disorder in Transactivation Regions of the NHRs

NHRs have two major transactivation domains, AF1 in the NTD, which acts in a constitutive manner, and a small but important AF2 towards the C-terminal end of the LBD, which functions in a ligand-dependent manner (Figure 10). AF2 is a binding surface created by a conformational change of helix 12 in the LBD after hormone binding (McEwan, 2012). AF2 preferentially interacts with short motifs rich in hydrophobic residues (FXXLF or LXXLL). Both transactivation domains can act autonomously, but there is increasing evidence that full receptor activity requires synergistic interactions between AF1 and AF2. In many cases, the synergistic effects are mediated by coregulatory proteins that may directly interact with both AF1 and AF2, albeit with different regions of the binding partner protein. Usually AF2 interacts with the LXXLL motif of the binding protein whereas no consensus motif has been identified for AF1 (Kumar and Litwack, 2009).

In the case of the androgen receptor (AR) there is direct interdomain cross-talk between the NTD and AF2. The first 20 residues of the AR-NTD contain a conserved pentameric sequence, FXXLF, which allows its interaction with AF2. X-ray crystallography of the LBD in complex with a peptide containing the FXXLF-motif showed that the peptide forms a two-turn amphipathic α-helix, where F23, L26, and F27 are directed toward the LBD surface and Q24, N25, and Q28 are exposed to the solvent (He et al., 2004). FRET and FRAP analysis showed that NTD-AF2 interaction occurs predominantly when ARs are mobile, possibly to prevent unfavorable or untimely cofactor interactions. This interaction is largely lost when AR transiently binds to DNA (van Royen et al., 2007).

2.2.1.1. Folding Induction Mediated by Protein-Protein Interactions

The NTD is involved in transcriptional activation and in protein-protein interactions. This region is negatively charged and it is the least conserved among the NHR's family members. It possesses powerful transactivation activity in those members with a long NTD sequence, as is the case of several steroid receptors (SRs) (Khorasanizadeh and Rastinejad, 2001).

AF1 is involved in protein-protein interactions with other transcription factors and the available data strongly suggest its coupled binding and folding. NMR and circular dichroism (CD) experiments have shown that the N-terminal domain of both estrogen receptors, ER-α and ER-β is disordered in solution (Warnmark et al., 2001). CD measurements showed that the complex of ERα with the TATA-binding protein (TBP) there was a decrease in the random coil component, suggesting the induction of folding in the NTD of the receptor upon binding In contrast, the N-terminal region of ERβ is ~80 amino acids shorter than that of ERα and has a low sequence identity with the other isoform.

FTIR and limited proteolysis have shown that the AF1 domain of the AR was also disordered (Kumar et al., 2004). Both ER and AR appear to have residual secondary structure. The AR AF1 domain can interact directly with the C-terminal domain of the RAP74 (RAP74-CTD) a subunit of the TFIIF complex. Examination of the complex between AR AF1 and RAP74 by FTIR showed an increase in the α-helical content from ~13% to ~40%, presumably due to the induction of secondary structure within the AF1 domain upon binding. This conformational change in AF1 enhanced subsequent binding of the p160 coactivator protein SRC-1a (Kumar et al., 2004). Mutation analysis has shown that when residues M244, L246 and V248 were replaced by alanines the receptor activity was reduced by 60% and the binding to RAP74 was abolished. However these mutations did not have a significant effect in the structure of the region or in the interaction with SRC-1a (Betney and McEwan, 2003).

2.2.1.2. Folding Induction by Osmolytes

Several factors can influence AF1 tertiary structure formation: binding of the DBD to DNA, ligand occupancy of the LBD, and in some circumstances, the type and concentrations of intracellular small solute molecules such as chemical chaperones or osmolytes (Kumar and Litwack, 2009). Several classes of well-known organic osmolytes are present in mammalian cells, where they function to protect proteins from denaturation under certain conditions; therefore it is possible that specific cells utilize specific osmolytes to encourage optimal functioning conformations of certain proteins. The presence of high osmolyte concentrations appeared to be important in certain cell-types where they counterbalance the effects of high contents of denaturants such as urea (Kumar et al., 2007).

The effects of the addition of three different types of osmolytes, amino acids, polyols and methylamines, in the induction of secondary structure in the transactivation domain AF1 of the glucocorticoid receptor (GR) were studied. The maximum of fluorescence emission of the recombinant AF1 (rAF1) in the presence of the three osmolytes showed a blue shift, suggesting the induction of a more ordered conformation. Furthermore, the conformational transition appeared to be cooperative. These results were supported by an increase in the resistance to proteolytic enzymes such as trypsin and chymotrypsin. The different osmolytes appeared to induce helical structure in rAF1 as judged by CD measurements. Functionally, folding of rAF1 facilitates its subsequent interaction with critical coregulatory proteins including TBP, CBP, and SRC-1 and enhance GR-induced transcription (Kumar et al., 2007; Kumar et al., 2004).

Similar results were obtained for the AF1 domain of the androgen receptor. In this case the presence of trimethylamine N-oxide (TMAO), a naturally occurring osmolyte induced the folding of the androgen receptor AF1 domain. This conformation appeared to enhance its binding to coregulators such as SCR-1 (Kumar et al., 2004). Furthermore, fluorescence spectroscopy results suggest that this region has features of molten-globule or pre-molten-globule-like states. This partially folded intermediate state may be important for the establishment of protein–protein interactions (Lavery and McEwan, 2006).

2.2.2. Intrinsic Disorder in DNA-Binding Regions

Two different domains of the NHRs participate in DNA-binding, the core DNA-binding domain (DBD) and the hinge domain. The DBD comprises ~66 residues in which conserved cysteine residues form two zinc-binding sites (Holmbeck et al., 1998b). High-resolution NMR studies of the RXR showed that residues 181-187 were not well ordered in the solution structure of the monomer (Figure 11A) (Holmbeck et al., 1998a). However, this region forms a distorted α-helix when homodimer or the heterodimer RXR-TR are bound to DNA (Figure 11B) (Rastinejad et al., 2000; Holmbeck et al., 1998a). Similar observations have been made for the second zinc finger regions of the ER and GR DBDs. The fact that this region is involved in the dimer interface formation led to the hypothesis that folding of this region upon DNA-binding stabilize an ordered protein conformation that would nucleate dimerization and hence lead to cooperativity (Holmbeck et al., 1998a; Holmbeck et al., 1998b).

The "hinge-domain" (H) or D-domain of the NHRs is poorly conserved at the primary amino acid sequence level, but shares common features among steroid receptors including an enrichment of basic residues and a nuclear localization signal (NLS) (Hill et al., 2012). This region is involved in DNA recognition, heterodimerization, coactivator/corepressor interactions and protein–protein interactions. This region is also called the C-terminal extension (CTE) of the DNA-binding domain as it participates in receptor DNA binding.

The CTE of steroids receptors contains a conserved doublet of basic amino acids R637-K638 (positions in the PR) that interacts with the minor groove flanking either side of the HRE (Hormone Recognition Element) half-sites. Mutation of this doublet in different members of the steroid receptors reduced the binding affinity of the receptor (Hill et al., 2012). Crystallographic studies have shown that the interaction with the minor groove appears to be mediated by different structures. In the case of the progesterone receptor the CTE forms an extended loop (Roemer et al., 2006), while in the glucocorticoid receptor it folds into a short α-helix (Meijsing et al., 2009).

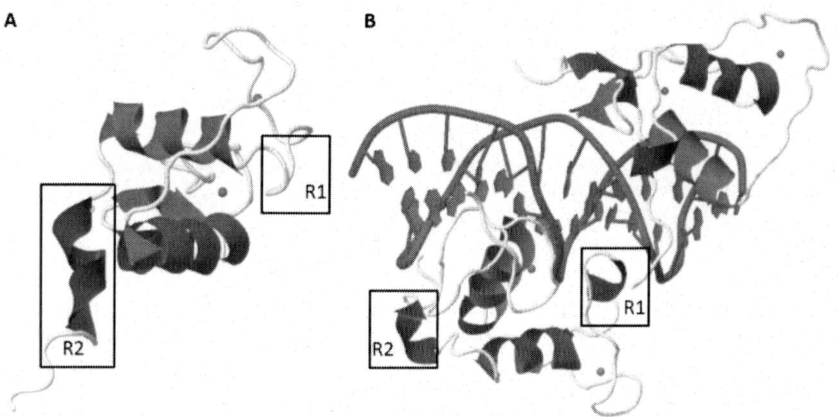

Figure 11. DNA induced conformational changes in RXR. A. Solution structure of the monomer RXR (PDB: 1RXR). B. Crystal structure of RXR homodimer in complex with DNA (PDB:1BY4). R1, corresponds to the region (181-187), which is unfolded in solution, but folds into a helical conformation upon DNA-binding. R2, corresponds to the region (202-209), that forms a stable helix in solution, which is disrupted after DNA-binding.

The second major group of NHRs includes non-steroid receptors that can bind DNA as heterodimers. The CTE of the receptors within group II forms a single α-helix, termed a T or A box, which can bind to DNA in different conformations. In the crystal structures of TR and VDR bound to DNA, the CTE forms a single α-helix that projects across the DNA minor groove between HRE half-sites and makes extensive non-specific contacts with the DNA backbone (Hill et al., 2012).

In the monomeric RXR receptor in solution the CTE is in α-helix conformation (Figure 11A) (Holmbeck et al., 1998a; Holmbeck et al., 1998b). The CTE α-helix spans from R202 to two or three residues beyond Q206. It is anchored to the protein by hydrophobic interactions between V205 and the core (Holmbeck et al., 1998a). Mutational analysis of this region showed that residues from 200 to 204 are important for DNA-binding, while residues beyond 204 participate in homodimer formation (Lee et al., 1993; Holmbeck et al., 1998a). This region unfolds to adopt an extended conformation in the homodimer and heterodimer complexes (i.e., RXR-RAR and RXR-TR) with DNA (Figure 5B) (Rastinejad et al., 2000; Zhao et al., 2000). X-ray crystallography of the homodimer bound to DNA showed that the unraveling of the helix in this region extends the backbone of the RXR bound to the 3' half-site so that residues at the C terminus can interact optimally with the upstream RXR DBD (Zhao et al., 2000). However, the α-helix disruption was observed in both components of the homodimer. When the C terminus is on the RXR subunit bound to the 5' half-site and therefore not part of the dimer interface, the loss of secondary structure may not contribute to cooperativity but may offset entropic costs of interaction and enhance protein-DNA binding (Holmbeck et al., 1998b).

While most nuclear hormone receptors bind DNA as homodimers or heterodimers, the estrogen-related receptors (ERRs) and other receptors such as NGFI-B and RevErb44 bind to single DNA half-sites as monomers (Dyson, 2012). These receptors contain a conserved sequence called GRIP-box (VRXGRIPK, where X is a Phe, Arg or Gly), which makes minor groove contacts with specific 5'flanking sequences of each HRE half-site (Zhao et al., 1998). The CTE lies along the floor of the minor groove in an extended loop conformation, providing additional specificity and affinity to the monomeric DNA-binding. Furthermore, in the case of ERR2 it has been shown that the CTE undergoes a conformational change from disordered in the free protein to well-ordered upon DNA binding (Gearhart et al., 2003).

2.2.3. PTMs in the Intrinsically Disordered Regions of NHRs

Post-translational modifications in the NHRs are mainly located in the intrinsically disordered regions: the NTD and the hinge domain. NTDs have been shown to be the site of both phosphorylation and sumoylation, whereas the hinge region gathers the sites of acetylation and ubiquitination (McEwan, 2004). PTMs modulate protein-protein interactions, DNA binding, and ligand binding, thereby altering gene expression, cell growth and differentiation (Blind and Garabedian, 2008).

Among the PTMs described for NHRs, site-specific phosphorylation appears to mediate conformational changes within GR AF1, protein-protein interactions and transactivation (Kumar and Thompson, 2012). Three major functionally important GR phosphorylation sites are conserved among humans (S203, S211, and S226), mouse, and rat, which are located within the AF1 domain (Chen et al., 2008). Phosphorylation of S203 and S211 has been shown to be hormone-dependent. GR S211 is reported to be a substrate for p38 mitogen-activated protein kinase (MAPK), suggesting a role for p38 MAPK signaling in glucocorticoid-induced apoptosis of lymphoid cells.

Circular dichroism combined with limited proteolysis and fluorescence emission spectroscopy have shown that phosphorylation of GR at S211 by p38 kinase induces secondary/tertiary structure in the AF1 domain. These results suggest that under physiological conditions, site-specific phosphorylation may play a crucial role in allowing the intrinsically disordered AF1 domain of the GR to adopt a functionally active conformation (Garza et al., 2010). Immunoprecipitation and FRET analysis showed that phosphorylation-induced structure formation in the GR AF1 facilitates its interaction with specific coregulatory proteins such as CBP, TBP, and SRC-1. Moreover, these interactions enhanced the transcriptional activity of the GR (Garza et al., 2010).

3. Intrinsically Disordered Proteins in DNA Metabolism

Intrinsic disorder is also present in DNA-binding proteins associated with DNA metabolic processes including replication, recombination and repair and chromatin modification. Table 1 contains a summary of this type of IDPs currently listed in the Disprot Database (Sickmeier et al., 2007). Although intrinsic disorder is more abundant in multicellular eukaryotes, this group contains bacterial and viral proteins as well. These proteins are highly heterogeneous, differing in number of IDRs, total percentages of disorder and specific functions. The role of intrinsic disorder in one member of this group, *Escherichia coli* single-stranded DNA-binding protein, is described.

Table 1. Intrinsically disordered proteins associated with DNA metabolic processes

Disprot Code	Protein name	Species	IDR Localization	Percentage of disorder (%)
DP00060	Modification methylase Pvull	Proteus vulgaris	70-81, 166 - 203	15%
DP00061	Replication protein A 70 kDa	Homo sapiens	115 - 168	9%
DP00152	DNA repair protein XRCC4	Homo sapiens	77-82 179-265	28%
DP00533	Regulatory protein SIR3 (Silent information regulator 3)	Saccharomyces cerevisiae	216-550	34%
DP00651_C007	Integrase p46	Moloney murine (MoMLV)	207-218	3%
DP00691	Regulator of chromosome condensation	Homo sapiens	2-20 198-223	12%
DP00711	Lysine-specific demethylase 5B	Mus musculus	115-128 159-164 178-208	3%
DP00712	Lysine-specific demethylase 5B	Homo sapiens	115-129 159-165 188-199	2%
DP00719	G/T mismatch-specific thymine DNA glycosylase (TDG)	Homo sapiens	1-111 340-410	44%
DP00721	FACT complex subunit spt16	Drosophila melanogaster	889 - 1044	14%
DP00722	Single-stranded DNA-binding protein	Escherichia coli	114-178	37%

IDR: Intrinsically disordered region

3.1. Single-Stranded DNA-Binding Proteins (SSBs)

Single-stranded DNA-binding proteins (SSBs) bind with high affinity in a sequence -independent manner to single stranded DNA (ssDNA). They protect ssDNA from degradation and recruit other proteins required for DNA replication, recombination, and repair. SSBs are present throughout all of life kingdoms and are indispensable for cell survival (Marceau, 2012).

The SSB proteins from eukaryotes, bacteria and bacteriophages have little sequence similarity, different subunit composition and oligomerization states. However, most SSBs contain at least one OB-fold (oligonucleotide/ oligosaccharide binding) DNA binding domain. The OB-fold domain is responsible for ssDNA binding and consists of five stranded β-sheets arranged as a β-barrel capped by a single α-helix. Its primary sequence is not well conserved, and the loops connecting the β-strands can vary in length and organization (Murzin et al., 1993)

The quaternary structure of the SSBs proteins and the mechanism of binding to ssDNA vary in bacteria, eukaryotes and bacteriophages. Many phages and viral single stranded binding proteins can function either as monomers (e.g., T4 gp32) or as dimers (e.g., T7 gp2.5) (Shamoo et al., 1995; Hollis et al., 2001). Bacteriophage SSB proteins can bind ssDNA in a highly cooperative manner, which leads to clustering of SSB protein to form protein filaments on long ssDNA (Shamon et al., 1995; Hollis et al., 2001; Suau et al., 1993).

The major eukaryotic SSB proteins are also called Replication Protein A, or RPA, and generally function as heterotrimers (RPA70, RPA34, and RPA14 subunits). They contain a total of six OB folds in its three subunits (Bochkarev et al., 1997). RPA is involved in many aspects of DNA metabolism such as DNA replication, recombination and repair. RPA70 interacts with several proteins including p53. In contrast to bacterial SSB, eukaryotic RPA does not display significant cooperativity in its binding to ssDNA (Bochkareva et al., 2001). RPA70 contains an intrinsically unstructured linker domain that separates a protein interaction and weak ssDNA binding domain (DBD F) from two high affinity ssDNA binding domains (DBD A and B) (Jacobs et al., 1999).

The bacterial SSB functions as homotetramers and contain a total of four OB folds (Figure 12). In the homotetramer each OB fold domain is able to bind to ssDNA. DNA wraps around the outside of the oligomeric protein (Shereda et al., 2008). The SSB from *Escherichia coli* (Ec-SSB) is the best studied of these proteins. Each subunit, (177 aa) is composed of two domains: the OB-fold N-terminal domain (112 aa) and an intrinsically disordered C-terminal domain (65 aa). The latter is composed of a flexible, intrinsically disordered linker (IDL) and a conserved motif called acid TIP (Figure 12A). The acid TIP consists in the last nine amino acids (MDFDDDIPF in *E. coli*) and provides the site for interaction with at least a dozen other proteins that function in DNA metabolism.

The DNA binding domain (OB fold) and the acid TIP are essential for *E. coli* survival, but the function or properties of the 56-aa linker that connects the OB fold and the TIP has been less investigated. The crystal structure of Ec-SSB bound to DNA (Savvides et al., 2004) revealed that the entire C-terminal domain is disordered even when bound to ssDNA. The absence of observable interactions with the core protein and the crystal packing suggest that the disordered C-terminal domains likely extend away from the DNA-binding domains, which may facilitate interactions with other proteins related to DNA replication, recombination, and repair.

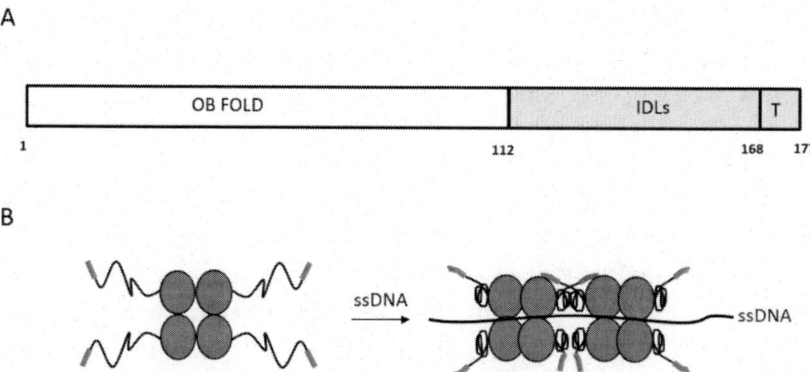

Figure 12. Structure of *E.coli* SSB. A. Structural domains of *E. coli* SSB (EcSSB) (Uniprot, P0AGE0), N-terminal DNA binding domain, OB fold (1–112 residues) and a C-terminal domain: Intrinsically disordered linker IDLs (56 aa) and a 9-residue acidic "tip" (T). In light gray, intrinsically disordered regions. B. Model for highly cooperative ssDNA binding in the SSB 35 mode (adapted of Kozlov 2015). Binding of SSB to ssDNA in its SSB 35 binding mode leaves DNA binding sites unoccupied in two subunits. Negatively charged tips of two C-terminal tails bind to the unoccupied subunits of adjacent tetramers. Positive cooperativity is enhanced due to inter-chain interactions between globular IDLs of adjacent tetramers. N-terminal DNA binding domain, OB fold: gray circle, IDLs : black thin line and a 9-residue acidic "tip": gray thick line.

Ec-SSB binds ssDNA in two distinct modes, which are named by the number of nucleotides occluded by the tetramer: SSB 65 and SSB 35. The stability of each binding mode is influenced *in vitro* by the concentration of monovalent salts, Mg $^{2+}$, spermidine and spermine. In the SSB 65 mode, all four OB domains in each tetramer interact with ssDNA with limited cooperativity (Lohman et al., 1994). The SSB 65 mode is favored under high ionic conditions (>200 mM NaCl) and low SSB:ssDNA ratios and has been proposed to function in homologous recombination. In the SSB 35 mode, two OB folds per SSB tetramer interact with the ssDNA. This binding mode is highly cooperative, stimulating SSB to form long nucleoprotein filaments that coat ssDNA (Lohman et al., 1994). This binding mode is more stable under low ionic conditions and high SSB:ssDNA ratios and has been proposed to function in DNA replication (Marceau, 2012).

Recently, it has been reported that the C-terminal domain is involved in the highly cooperative binding of Ec-SSB to ssDNA in its SSB 35 mode. Cooperativity is affected by the length and composition of IDLs and by the 9-aa tip. Complete deletion of the linker eliminates the highly cooperative binding of SSB to ssDNA *in vitro*, while the removal of the acid TIP only reduced cooperative binding (Kozlov et al., 2015).

Extremophilic bacteria such as *Deinococcus radiodurans* and *Thermus aquaticus* have a dimeric version of SSB, in which each subunit contains two OB-folds; hence, the DNA binding core still possesses four OB-folds and thus is structurally similar to the homotetrameric SSB (Kelman et al., 1998; Fedorov et al., 2006). Comparison of the crystal structures and DNA binding properties of Dr-SSB (*Deinococcus radiodurans* SSB) and Ec-SSB suggest that they share similar mechanisms of DNA binding and wrapping (George et al., 2012). In Dr-SSB, two C-terminal tails by tetramer are enough to mediate protein–protein interactions.

Studies with mutated Ec-SSB variants have shown that the two-tailed SSB "tetramer" is functional *in vivo* and is competent for DNA replication *in vitro*, but shows defects in DNA repair. The two-tailed SSB tetramer also retained cooperative binding to ssDNA. However, a single-tailed SSB "tetramer" could not carry out one or more essential functions *in vivo* and showed a reduction in cooperative binding (Antony et al., 2013). These results suggest that of the four C-terminal tails present in the wild-type Ec-SSB, only two are involved in SSB cooperative binding, while the other two are free to interact with other proteins (Kozlov et al., 2015).

Most IDPs are polyampholytes, with sequences that include both positively and negatively charged residues. The fraction of charged residues discriminates between weak and strong polyampholytes. Recent studies, using atomistic simulations, have shown that weak polyampholytes form globules, whereas the conformational preferences of strong polyampholytes are determined by a combination of the fraction of charged residues and the linear sequence distributions of oppositely charged residues (Das et al., 2013).

Analysis of the amino acid composition of the majority the intrinsically disordered linkers of bacterial SSBs revealed that they are weak polyampholytes and can be classified as globule formers. Atomistic simulations using the sequence of the IDL of Ec-SSB suggested that this region forms a heterogeneous distribution of globules (Kozlov et al., 2015). Hydrodynamic measurements by analytical sedimentation are in agreement with the formation of a globular structure by the IDL. Variations in the linker lengths did not induced significant changes in the overall dimensions of the globule. This would explain why Ec-SSB constructs with shorter IDLs still display highly cooperative binding to ssDNA (Kozlov et al., 2015).

Based in these results, Kozlov et al. (2015) proposed a model to explain the role of the intrinsically disordered linker (IDL) of Ec-SSB in the SSB 35 binding mode. In this model the formation of a globular structure by the IDL promotes intra-tetramer and inter-tetramer interactions, which in turn, results in the highly cooperative binding of Ec-SSB to ssDNA (Figure 12B).

CONCLUSION

Intrinsic disorder is common in DNA-binding proteins. This highly heterogeneous group includes completely disordered proteins like HMGA or protamines and also partially disordered proteins. Intrinsic disorder can be found in DNA-binding domains as well as in other domains involved in protein-protein interactions. The advantages of intrinsic disorder in DNA-binding domains are associated with a larger capture radius and an increased speed of DNA search. Intrinsically disordered domains involved in protein-protein interactions may act as protein hubs with multiple partners interacting with the same region, as is the case of the transcriptional activation region of TFs. Most of the DNA-binding IDPs undergo coupled binding and folding into well-defined structures, but in some cases they can also form fuzzy complexes after binding to their targets. Induced structures may be specific according to the template provided by structured partners and modulated by post-translational modifications.

REFERENCES

Adams, VH; McBryant, SJ; Wade, PA; Woodcock, CL; Hansen, JC. Intrinsic disorder and autonomous domain function in the multifunctional nuclear protein, MeCP2. *J. Biol. Chem.*, 2007 282, 15057–15064.

Allan, J; Hartman, PG; Crane-Robinson, C; Aviles, FX. The structure of histone H1 and its location in chromatin. *Nature*, 1980 288(5792), 675–679.

Allan, J; Mitchell, T; Harborne, N; Bohm, L; Crane-Robinson, C. Roles of H1 domains in determining higher order chromatin structure and H1 location. *J. Mol. Biol.*, 1986 187(4), 591–601.

Ali T, Coles P, Stevens TJ, Stott K, Thomas JO. Two homologous domains of similar structure but different stability in the yeast linker histone, Hho1p. J Mol Biol. 2004 Apr 16;338(1):139-48.

Antony, E; Weiland, E; Yuan, Q; Manhart, CM; Nguyen, B; Kozlov, AG; McHenry, CS; Lohman TM. Multiple C-Terminal Tails within a Single E. coli SSB Homotetramer Coordinate DNA Replication and Repair *J. Mol. Biol.*, 2013 425(23), 4802-4819.

Arai, M; Sugase, K; Dyson, HJ; Wright, PE. Conformational propensities of intrinsically disordered proteins influence the mechanism of binding and folding. *Proc. Natl. Acad. Sci. USA.*, 2015 112(31), 9614-9619.

Banères, JL; Martin, A; Parello, J. The N tails of histones H3 and H4 adopt a highly structured conformation in the nucleosome. *J. Mol. Biol.*, 1997 273(3), 503-508.

Barlev, NA; Liu, L; Chehab, NH; Mansfield, K; Harris, KG; Halazonetis, TD; Berger, SL. Acetylation of p53 activates transcription through recruitment of coactivators/histone acetyltransferases. *Mol. Cell*, 2001 8, 1243–1254.

Betney, R; McEwan, IJ. Role of conserved hydrophobic amino acids in androgen receptor AF-1 function. *J. Mol. Endocrinol.*, 2003 31(3), 427-439.

Blind, RD; Garabedian, MJ; Differential recruitment of glucocorticoid receptor phospho-isoforms to glucocorticoid-induced genes. *J. Steroid Biochem. Mol. Biol.*, 2008 109, 150–157.

Bloch DP. A catalog of sperm histones. *Genetics,* 1969; 61(1), Suppl, 93-111.

Bochkarev, A; Pfuetzner, RA; Edwards, AM; Frappier, L. Structure of the single-stranded-DNA-binding domain of replication protein A bound to DNA. *Nature* 1997 385 (6612), 176–181.

Bochkareva, E; Belegu, V; Korolev, S; Bochkarev A. Structure of the major single-stranded DNA-binding domain of replication protein A suggests a dynamic mechanism for DNA binding. *EMBO J.*, 2001 20(3), 612–618.

Bochkareva, E; Kaustov, L; Ayed, A; Yi, GS; Lu, Y; Pineda-Lucena, A; Liao, JC; Okorokov, AL; Milner, J; Arrowsmith, CH; Bochkarev, A. Single-stranded DNA mimicry in the p53 transactivation domain interaction with replication protein A. *Proc. Natl. Acad. Sci. USA*, 2005 102(43), 15412-15417.

Böhm, L; Mitchell, TC. Sequence conservation in the N-terminal domain of histone H1. *FEBS Lett.*, 1985 193(1), 1-4.

Carruthers, LM; Hansen, JC. The core histone N termini function independently of linker histones during chromatin condensation. *J. Biol. Chem.*, 2000 275(47), 37285-37290.

Caterino, TL; Fang, H; Hayes, JJ. Nucleosome linker DNA contacts and induces specific folding of the intrinsically disordered H1 carboxyl-terminal domain. *Mol. Cell Biol.* 2011, 31(11), 2341-2348.

Caterino, TL; Hayes, JJ. Structure of the H1 C-terminal domain and function in chromatin condensation. *Biochem. Cell Biol.*, 2011 89(1), 35-44.

Caterino, TL; Hayes, JJ. Chromatin structure depends on what's in the nucleosome's pocket. *Nat. Struct. Mol. Biol.*, 2007 14(11), 1056-8.

Catez, F; Hock, R. Binding and interplay of HMG proteins on chromatin, lessons from live cell imaging. *Biochim. Biophys. Acta*, 2010 1799(1-2), 15-27.

Chapman, GE; Hartman, PG; Bradbury, EM. Studies on the role and mode of operation of the very-lysine-rich histone H1 in eukaryote chromatin. The isolation of the globular and non-globular regions of the histone H1 molecule. *Eur. J. Biochem.*, 1976 61(1), 69-75.

Chen, W; Dang, T; Blind, RD; Wang, Z; Cavasotto, CN; Hittelman, AB; Rogatsky, I; Logan, SK; Garabedian, MJ. Glucocorticoid receptor phosphorylation differentially affects target gene expression. *Mol. Endocrinol.*, 2008 22, 1754–1766.

Chi, SW; Lee, SH; Kim, DH; Ahn, MJ; Kim, JS; Woo, JY; Torizawa, T; Kainosho, M; Han, KH. Structural details on mdm2-p53 interaction. *J. Biol. Chem.*, 2005 280(46), 38795-38802.

Clark, DJ; Hill, CS; Martin, SR; Thomas, JO. Alpha-helix in the carboxy-terminal domains of histones H1 and H5. *EMBO J.*, 1988 7(1), 69-75.

Das, RK; Crick, SL; Pappu, RV. N-terminal segments modulate the α-helical propensities of the intrinsically disordered basic regions of bZIP proteins. J. Mol. Biol. 2012 416(2), 287-299.

Das, RK; Pappu, RV. Conformations of intrinsically disordered proteins are influenced by linear sequence distributions of oppositely charged residues. *Proc. Natl. Acad. Sci. USA*, 2013 110, 13392–13397.

Di Lello, P; Jenkins, LM; Jones, TN; Nguyen, BD; Hara, T; Yamaguchi, H; Dikeakos, JD; Appella, E; Legault, P; Omichinski, JG. Structure of the Tfb1/p53 complex, Insights into the interaction between the p62/Tfb1 subunit of TFIIH and the activation domain of p53. *Mol. Cell.*, 2006 22(6), 731-740.

Dou, Y; Gorovsky, MA. Phosphorylation of linker histone H1 regulates gene expression in vivo by creating a charge patch. *Mol. Cell.*, 2000 6(2), 225-231.

Dreveny, I; Deeves, SE; Fulton, J; Yue, B; Messmer, M; Bhattacharya, A; Collins, HM; Heery, DM. The double PHD finger domain of

MOZ/MYST3 induces α-helical structure of the histone H3 tail to facilitate acetylation and methylation sampling and modification. *Nucleic Acids Res.*, 2014 42(2), 822-835.

Dunker, AK; Bondos, SE; Huang, F; Oldfield, CJ. Intrinsically disordered proteins and multicellular organisms. *Semin. Cell Dev. Biol.*, 2015 37, 44-55.

Dyson HJ. Roles of intrinsic disorder in protein-nucleic acid interactions. *Mol. Biosyst.*, 2012 8(1), 97-104.

Eisert, RJ; Kennedy, SA; Waters, ML. Investigation of the β-sheet interactions between dHP1 chromodomain and histone 3. *Biochemistry,* 2015 54(14), 2314-2322.

Fang, H; Clark, DJ; Hayes, JJ. DNA and nucleosomes direct distinct folding of a linker histone H1 C-terminal domain. *Nucleic Acids Res.,* 2012 40(4), 1475-1484.

Fedorov, R; Witte, G; Urbanke, C; Manstein, DJ; Curth, U. 3D structure of Thermus aquaticus single-stranded DNA-binding protein gives insight into the functioning of SSB proteins. *Nucleic Acids Res.*, 2006 34, 6708–6717.

Feng, H; Jenkins, LM; Durell, SR; Hayashi, R; Mazur, SJ; Cherry, S; Tropea, JE; Miller, M; Wlodawer, A; Appella, E; Bai, Y. Structural basis for p300 Taz2-p53 TAD1 binding and modulation by phosphorylation. *Structure,* 2009 17(2), 202-210.

Fonfría-Subirós, E; Acosta-Reyes, F; Saperas, N; Pous, J; Subirana, JA; Campos, JL. Crystal structure of a complex of DNA with one AT-hook of HMGA1. *PLoS One,* 2012 7(5), e37120.

Frankel, AD; Berg, JM; Pabo, CO. Metal-dependent folding of a single zinc finger from transcription factor IIIA. *Proc. Natl. Acad. Sci. USA*, 1987 84(14), 4841-4845.

Ganguly, D; Chen, J. Modulation of the disordered conformational ensembles of the p53 transactivation domain by cancer-associated mutations. *PLoS Comput. Biol.*, 2015 11(4), e1004247.

Garza, AM; Khan, SH; Kumar, R. Site-specific phosphorylation induces functionally active conformation in the intrinsically disordered N-terminal activation function (AF1) domain of the glucocorticoid receptor. *Mol. Cell Biol.*, 2010 30(1), 220-230.

Gearhart, MD; Holmbeck, SM; Evans, RM; Dyson, HJ; Wright, PE. Monomeric complex of human orphan estrogen related receptor-2 with DNA, a pseudo-dimer interface mediates extended half-site recognition. *J. Mol. Biol.*, 2003 327(4), 819-32.

George, NP; Ngo, KV; Chitteni-Pattu, S; Norais, CA; Battista, JR; Cox, MM, et al. Structure and cellular dynamics of Deinococcus radiodurans single-stranded DNA (ssDNA)-binding protein (SSB)-DNA complexes *J. Biol. Chem.*, 2012 287, 22123–22132.

Gréen, A; Sarg, B; Gréen, H; Lönn, A; Lindner, H; Rundquist, I. Histone H1 interphase phosphorylation becomes largely established in G1 or early S phase and differs in G1 between T-lymphoblastoid cells and normal T cells. *Epigenetics Chromatin*, 2011 4, 15.

Hansen, JC; Lu, X; Ross, ED; Woody, RW. Intrinsic protein disorder, amino acid composition, and histone terminal domains. *J. Biol. Chem.*, 2006 281(4), 1853-1856.

Hansen, JC; Wexler, BB; Rogers, DJ; Hite, KC; Panchenko, T; Ajith, S; Black, BE. DNA binding restricts the intrinsic conformational flexibility of methyl CpG binding protein 2 (MeCP2*)*. *J. Biol. Chem.*, 2011 286(21), 18938-18948.

Harshman, SW; Young, NL; Parthun, MR; Freitas, MA. H1 histones, current perspectives and challenges. *Nucleic Acids Res.* 2013 41(21), 9593-9609.

Hayashi, T; Hayashi, H; Iwai, K. Tetrahymena histone H1. Isolation and amino acid sequence lacking the central hydrophobic domain conserved in other H1 histones *J Biochem.* 1987 102(2), 369-376.

He, B; Gampe, RT Jr; Kole, AJ; Hnat, AT; Stanley, TB; An, G; Stewart, EL; Kalman, RI; Minges, JT; Wilson, EM. Structural basis for androgen receptor interdomain and coactivator interactions suggests a transition in nuclear receptor activation function dominance. *Mol. Cell.*, 2004 16(3), 425-438.

Hendzel, MJ; Lever, MA; Crawford, E; Th'ng, JP. The C-terminal domain is the primary determinant of histone H1 binding to chromatin in vivo. *J. Biol. Chem.*, 2004 279(19), 20028-20034.

Hill, KK; Roemer, SC; Churchill, ME; Edwards, DP. Structural and functional analysis of domains of the progesterone receptor. *Mol. Cell Endocrinol.*, 2012 348(2), 418-429.

Hite, KC; Adams, VH; Hansen, JC. Recent advances in MeCP2 structure and function. *Biochem. Cell Biol.*, 2009 87(1), 219-227.

Hite, KC; Kalashnikova, AA; Hansen, JC. Coil-to-helix transitions in intrinsically disordered methyl CpG binding protein 2 and its isolated domains. *Protein Sci.*, 2012 21(4), 531-5238.

Ho, KL; McNae, IW; Schmiedeberg, L; Klose, RJ; Bird, AP; Walkinshaw, MD. MeCP2 binding to DNA depends upon hydration at methyl-CpG. *Mol. Cell*, 2008 29, 525–531.

Hock, R; Furusawa,T; Ueda,T; Bustin, M. HMG chromosomal proteins in development and disease. *Trends Cell Biol.*, 2007 17(2), 72-79.

Hoff, KG; Avalos, JL; Sens, K; Wolberger, C. Insights into the sirtuin mechanism from ternary complexes containing NAD+ and acetylated peptide. *Structure,* 2006 14(8), 1231-1240.

Hollis, T; Stattel, JM; Walther, DS; Richardson, CC; Ellenberger, T. Structure of the gene 2.5 protein, a single-stranded DNA binding protein encoded by bacteriophage T7. *Proc. Natl. Acad. Sci. USA*, (2001) 98(17), 9557–9562

Holmbeck, SM; Dyson, HJ; Wright, PE. DNA-induced conformational changes are the basis for cooperative dimerization by the DNA binding domain of the retinoid X receptor. *J. Mol. Biol.*, 1998 284(3), 533-539.

Holmbeck, SM; Foster, MP; Casimiro, DR; Sem, DS; Dyson, HJ; Wright, PE. High-resolution solution structure of the retinoid X receptor DNA-binding domain. *J. Mol. Biol.*, 1998 281(2), 271-284.

Hollenbeck, JJ; McClain, DL; Oakley, MG. The role of helix stabilizing residues in GCN4 basic region folding and DNA binding. *Protein Sci.*, 2002 11, 2740-2747.

Hughes, RM; Wiggins, KR; Khorasanizadeh, S; Waters, ML. Recognition of trimethyllysine by a chromodomain is not driven by the hydrophobic effect. *Proc. Natl. Acad. Sci. USA.*, 2007 104(27), 11184-11188.

Hupp, TR; Meek, DW; Midgley, CA; Lane, DP. Regulation of the specific DNA binding function of p53. *Cell*, 1992 71(5), 875-886.

Hutchinson, JB; Cheema, MS; Wang, J; Missiaen, K; Finn, R; Gonzalez-Romero, R; Th'ng, JP; Hendzel, M; Ausio, J. Interaction of chromatin with a histone H1 containing swapped N- and C-terminal domains. *Biosci. Rep.*, 2015 35(3), pii, e00209.

Huth, JR; Bewley, CA; Nissen, MS; Evans, JN; Reeves, R; Gronenborn, AM; Clore, GM. The solution structure of an HMG-I(Y)-DNA complex defines a new architectural minor groove binding motif. *Nat. Struct. Biol.*, 1997 4(8), 657-665.

Jacobs, DM; Lipton, AS; Isern, NG; Daughdrill, GW, Lowry DF, Gomes X, Wold MS. Human replication protein A, global fold of the N-terminal RPA-70 domain reveals a basic cleft and flexible C-terminal linker. *J. Biomol. NMR*, 1999 14(4), 321-31.

Jacobs, SA; Khorasanizadeh, S. Structure of HP1 chromodomain bound to a lysine 9-methylated histone H3 tail. *Science*, 2002 295(5562), 2080-2083.

Jenkins, LM; Yamaguchi, H; Hayashi, R; Cherry, S; Tropea, JE; Miller, M; Wlodawer, A; Appella, E; and Mazur, M.J. Two distinct motifs within the p53 transactivation domain bind to the Taz2 domain of p300 and are differentially affected by phosphorylation. *Biochemistry*, 2009 48(6), 1244-1255.

Kalashnikova, AA; Porter-Goff, ME; Muthurajan, UM; Luger, K; Hansen, JC. The role of the nucleosome acidic patch in modulating higher order chromatin structure. *J. R. Soc. Interface*, 2013 10(82), 20121022.

Kan, PY; Lu, X; Hansen, JC; Hayes, JJ. The H3 tail domain participates in multiple interactions during folding and self-association of nucleosome arrays. *Mol. Cell Biol.*, 2007 27(6), 2084-2091.

Kaustov, L; Ouyang, H; Amaya, M; Lemak, A; Nady, N; Duan, S; Wasney, GA; Li, Z; Vedadi, M; Schapira, M; Min, J; Arrowsmith, CH. Recognition and specificity determinants of the human cbx chromodomains. *J. Biol. Chem.*, 2011 286(1), 521-529.

Kelman, Z; Yuzhakov, A; Andjelkovic, J; O'Donnell, M. Devoted to the lagging strand-the subunit of DNA polymerase III holoenzyme contacts SSB to promote processive elongation and sliding clamp assembly. *EMBO J.*, 1998 17, 2436–2449.

Khazanov, N; Levy, Y. Sliding of p53 along DNA can be modulated by its oligomeric state and by cross-talks between its constituent domains. *J. Mol. Biol.* 2011 408(2), 335-355.

Khorasanizadeh, S; Rastinejad, F. Nuclear-receptor interactions on DNA-response elements. *Trends Biochem Sci.*, 2001 26(6), 384-390.

Kostova, NN; Srebreva, L; Markov, DV; Sarg, B; Lindner, HH; Rundquist, I. Histone H5-chromatin interactions in situ are strongly modulated by H5 C-terminal phosphorylation. *Cytometry A*, 2013 83(3), 273-279.

Kozlov, AG; Weiland, E; Mittal, A; Waldman, V; Antony, E; Fazio, N; Pappu, RV; Lohman TM. Intrinsically disordered C-terminal tails of E. coli single-stranded DNA binding protein regulate cooperative binding to single-stranded DNA. *J. Mol. Biol.*, 2015 427(4), 763-774.

Krasowski, MD; Reschly, EJ; Ekins, S. Intrinsic disorder in nuclear hormone receptors. *J. Proteome Res.*, 2008 (10), 4359-4372.

Krishna, SS; Majumdar, I; Grishin, NV. Structural classification of zinc fingers, survey and summary. *Nucleic Acids Research*, 2003 31(2), 532–550.

Kumar, R; Betney, R; Li, J; Thompson, EB; McEwan, IJ. Induced alpha-helix structure in AF1 of the androgen receptor upon binding transcription factor TFIIF. *Biochemistry*, 2004 43, 3008–3013.

Kumar, R; Litwack, G. Structural and functional relationships of the steroid hormone receptors' N-terminal transactivation domain. *Steroids*, 2009 74(12), 877-83.

Kumar, R; Serrette, JM; Khan, SH; Miller, AL; Thompson, EB. Effects of different osmolytes on the induced folding of the N-terminal activation domain (AF1) of the glucocorticoid receptor. *Arch. Biochem. Biophys.*, 2007 465(2), 452-460.

Kumar, R; Thompson, EB. Folding of the glucocorticoid receptor N-terminal transactivation function, dynamics and regulation. *Mol. Cell Endocrinol.*, 2012 348(2), 450-456.

Kussie, PH; Gorina, S; Marechal, V; Elenbaas, B; Moreau, J; Levine, AJ; Pavletich, NP. Structure of the MDM2 oncoprotein bound to the p53 tumor suppressor transactivation domain. *Science,* 1996 274(5289), 948-53.

Lavery, DN; McEwan, IJ. The human androgen receptor AF1 transactivation domain, interactions with transcription factor IIF and molten-globule-like structural characteristics. *Biochem. Soc. Trans.*, 2006 34, 1054-1057.

Lee, CW; Arai, M; Martinez-Yamout, MA; Dyson, HJ; Wright PE. Mapping the interactions of the p53 transactivation domain with the KIX domain of CBP. *Biochemistry*, 2009 48(10), 2115-2124.

Lee, CW; Martinez-Yamout, MA; Dyson, HJ; Wright, PE. Structure of the p53 transactivation domain in complex with the nuclear receptor coactivator binding domain of CREB binding protein. *Biochemistry*, 2010a 49(46), 9964-9971.

Lee, CW; Ferreon, JC; Ferreon, AC; Arai, M; Wright, PE. Graded enhancement of p53 binding to CREB-binding protein (CBP) by multisite phosphorylation. *Proc. Natl. Acad. Sci. USA,* 2010b 107(45), 19290-19295.

Lee, H; Mok, KH; Muhandiram, R; Park, KH; Suk, JE; Kim, DH; Chang, J; Sung, YC; Choi, KY; Han, KH. Local structural elements in the mostly unstructured transcriptional activation domain of human p53. *J. Biol. Chem.*, 2000 275(38), 29426-2932.

Lee, JT; Gu, W. SIRT1, Regulator of p53 Deacetylation. *Genes Cancer.*, 2013 4(3-4), 112-117.

Lee, MS; Kliewer, SA; Provencal, J; Wright, PE; Evans, RM. Structure of the retinoid X receptor a DNA-binding domain, a helix required for homodimeric DNA-binding. *Science,* 1993 260, 1117-1121.

Lever, MA; Th'ng, JP; Sun, X; Hendzel, MJ. Rapid exchange of histone H1.1 on chromatin in living human cells. *Nature,* 2000 408(6814), 873-876.

Levine, AJ; Oren M. The first 30 years of p53, growing ever more complex. *Nat. Rev. Cancer*, 2009 9(10), 749-758.

Liu, J; Perumal, NB; Oldfield, CJ; Su, EW; Uversky, VN; Dunker, AK. Intrinsic disorder in transcription factors. *Biochemistry*. 2006 45(22), 6873-88.

Lohman, TM; Ferrari, ME. Escherichia coli single-stranded DNA-binding protein, multiple DNA-binding modes and cooperativities. *Annu Rev Biochem*, 1994 63, 527–570.

Lopez, R; Sarg, B; Lindner, H; Bartolomé, S; Ponte, I; Suau, P; Roque, A. Linker histone partial phosphorylation, effects on secondary structure and chromatin condensation. *Nucleic Acids Res.*, 2015 43(9), 4463-4476.

Lu, X; Hansen, JC. Identification of specific functional subdomains within the linker histone H10 C-terminal domain. *J. Biol. Chem.*, 2004 279(10), 8701-8707.

Manning GS. Is a small number of charge neutralizations sufficient to bend nucleosome core DNA onto its superhelical ramp? *J. Am. Chem. Soc.* 2003, 125(49), 15087-15092.

Marceau, AH. Functions of single-strand DNA-binding proteins in DNA replication, recombination, and repair. *Methods Mol. Biol.*, 2012 922, 1-21.

McBryant, SJ; Adams, VH; Hansen, JC. Chromatin architectural proteins. *Chromosome Res.*, 2006 14(1), 39-51.

McBryant, SJ; Klonoski, J; Sorensen, TC; Norskog, SS; Williams, S; Resch, MG; Toombs, JA 3rd; Hobdey, SE; Hansen, JC. Determinants of histone H4 N-terminal domain function during nucleosomal array oligomerization, roles of amino acid sequence, domain length, and charge density. *J. Biol. Chem.*, 2009 284(25), 16716-16722.

McDowell, C; Chen, J; Chen, J. Potential conformational heterogeneity of p53 bound to S100B(ββ). *J. Mol. Biol.*, 2013 425(6), 999-1010.

McEwan IJ. Intrinsic disorder in the androgen receptor, identification, characterisation and drugability. *Mol. Biosyst.*, 2012 8(1), 82-90.

McEwan, IJ. Sex, drugs and gene expression, signalling by members of the nuclear receptor superfamily. *Essays Biochem.*, 2004 40, 1-10.

McKinney, K; Mattia, M; Gottifredi, V; Prives, C. p53 linear diffusion along DNA requires its C terminus. *Mol. Cell*, 16(3), 413-424.

Meijsing, SH; Pufall, MA; So, AY; Bates, DL; Chen, L; Yamamoto, KR. DNA binding site sequence directs glucocorticoid receptor structure and activity. *Science*, 2009 324, 407–410.

Mujtaba, S; He, Y; Zeng L; Yan, S; Plotnikova, O; Sachchidanand, S; Sanchez, R; Zeleznik-Le NJ; Ronai, Z; Zhou, MM. Structural mechanism of the bromodomain of the coactivator CBP in p53 transcriptional activation. *Mol. Cell*, 2004 13(2), 251-263.

Murzin; AG. OB(oligonucleotide/oligosaccharide binding)-fold, common structural and functional solution for nonhomologous sequences. *EMBO J.* 1993 12(3), 861–867.

Öberg, C; Belikov, S. The N-terminal domain determines the affinity and specificity of H1 binding to chromatin. *Biochem. Biophys. Res. Commun.* 2012 420 (2), 321-324.

Orrego, M; Ponte, I; Roque, A; Buschati, N; Mora, X; Suau, P. Differential affinity of mammalian histone H1 somatic subtypes for DNA and chromatin. *BMC Biol.*, 2007 5, 22.

Piskacek, M; Vasku, A; Hajek, R; Knight, A. Shared structural features of the 9aaTAD family in complex with CBP. *Mol. Biosyst.*, 2015 11(3), 844-851.

Polley, S; Guha, S; Roy, NS; Kar, S; Sakaguchi, K; Chuman, Y; Swaminathan, V; Kundu, T; and Roy, S. Differential recognition of phosphorylated transactivation domains of p53 by different p300 domains. *J. Mol. Biol.*, 2008 376, 8–12.

Ponte, I, Vidal-Taboada, JM; Suau, P. Evolution of the vertebrate H1 histone class, evidence for the functional differentiation of the subtypes. *Mol. Biol. Evol.*, 1998 15(6), 702-708.

Postnikov, YV; Bustin, M. Functional interplay between histone H1 and HMG proteins in chromatin. *Biochim. Biophys. Acta.*, 2015 pii:S1874-9399(15)00212-6.

Raghuram, N; Strickfaden, H; D. McDonald, D; Williams, K; Fang, H; Mizzen, C; Hayes, JJ; Th'ng, JP, Hendzel, MJ. Pin1 promotes histone H1 dephosphorylation and stabilizes its binding to chromatin. *J. Cell Biol.*, 2013 203(1), 57-71.

Rastinejad, F; Wagner, T; Zhao, Q; Khorasanizadeh, S. Structure of the RXR-RAR DNA-binding complex on the retinoic acid response element DR1. *EMBO J.*, 2000 19(5), 1045-1054.

Reed, SM; Quelle, DE. *p53 Acetylation, Regulation and Consequences. Cancers*, 2014 7(1), 30-69.

Reeves, R. Nuclear functions of the HMG proteins. *Biochim. Biophys. Acta.*, 2010 1799(1-2), 3-14.

Roemer, SC; Donham, DC; Sherman, L; Pon, VH; Edwards, DP; Churchill, ME. Structure of the progesterone receptor-deoxyribonucleic acid complex, novel interactions required for binding to half-site response elements. *Mol. Endocrinol.*, 2006 20(12), 3042-3052.

Roque, A; Iloro, I. Ponte, I; Arrondo, JL; Suau, P. DNA induced secondary structure of the carboxyl-terminal domain of histone H1. *J. Biol. Chem.*, 2005 280(37), 32141–32147.

Roque, A; Ponte, I; Arrondo, JL; Suau, P. Phosphorylation of the carboxy-terminal domain of histone H1, effects on secondary structure and DNA condensation. *Nucleic Acids Res.*, 2008 36(14), 4719-4726.

Roque, A; Ponte, I; Suau, P, Macromolecular crowding induces a molten globule state in the C-terminal domain of histone H1. *Biophys J.*, 2007 93(6), 2170-2177.

Roque, A; Ponte, I; Suau, P. Role of charge neutralization in the folding of the carboxy-terminal domain of histone H1. *J. Phys. Chem. B.*, 2009 113, 3512061-3512066.

Roque, A; Ponte, I; Suau, P. Secondary structure of protamine in sperm nuclei, an infrared spectroscopy study.*BMC Struct. Biol.*, 2011 11, 14.

Roque, A; Teruel, N; Lopez, R; Ponte, I; Suau, P. Contribution of hydrophobic interactions to the folding and fibrillation of histone H1 and its carboxy-terminal domain. *J. Struct. Biol.*, 2012 180(1), 101-109.

Roth, SY; Allis, CD. Chromatin condensation, does histone H1 dephosphorylation play a role? *Trends Biochem. Sci.*, 1992 17(3), 93-98.

Roy, S; Musselman, CA; Kachirskaia, I; Hayashi, R; Glass, KC; Nix, JC; Gozani, O; Appella, E; Kutateladze, TG. Structural insight into p53 recognition by the 53BP1 tandem Tudor domain. *J. Mol. Biol.*, 2010 398(4), 489-496.

Rustandi RR; Baldisseri DM; Weber DJ. Structure of the negative regulatory domain of p53 bound to S100B(betabeta). *Nat. Struct. Biol.*, 2000 7(7), 570-574.

Sancho, M; Diani, E; Beato, M; Jordan; A. Depletion of human histone H1 variants uncovers specific roles in gene expression and cell growth. *PLoS Genet.*, 2008 4(10), e1000227.

Savvides, SN; Raghunathan, S; Fütterer, K; Kozlov, AG; Lohman, TM; Waksman, G. The C-terminal domain of full-length E. coli SSB is disordered even when bound to DNA. *Protein Sci.*, 2004 13(7), 1942-1947.

Schumacher, MA; Goodman, RH; Brennan, RG. The structure of a CREB bZIP.somatostatin CRE complex reveals the basis for selective dimerization and divalent cation-enhanced DNA binding. *J. Biol. Chem.*, 2000 275, 35242-35247.

Serrano, MA; Li, Z; Dangeti, M; Musich, PR; Patrick, S; Roginskaya, M; Cartwright, B; Zou Y. DNA-PK, ATM and ATR collaboratively regulate p53-RPA interaction to facilitate homologous recombination DNA repair. *Oncogene,* 2013 32(19), 2452-62.

Shamoo, Y; Friedman, AM; Parsons, MR; Konigsberg, WH; Steitz, TA. Crystal structure of a replication fork single-stranded DNA binding protein (T4 gp32) complexed to DNA. *Nature,* 1995 376(6538), 362–366.

Shereda, RD; Kozlov, AG; Lohman, TM; Cox, MM; Keck, JL. SSB as an organizer/mobilizer of genome maintenance complexe. *Crit. Rev. Biochem. Mol. Biol.,* 2008 43(5), 289-318.

Sickmeier, M; Hamilton, JA; LeGall, T; Vacic, V; Cortese, MS; Tantos, A; Szab, B; Tomp, P; Chen, J; Uversky, VN; Obradovic, Z; Dunker, AK. DisProt, the Database of Disordered Proteins. *Nucleic Acids Res.,* 2007 35(Database issue), D786-93.

Suau P; Toulmé JJ; Hélène C. The binding of T4 gene 32 protein to MS2 virus RNA and transfer RNA. *Nucleic Acids Res.,* 1980 8(6), 1357-1372.

Subirana, JA; Puigjaner L. X-ray diffraction studies of nucleohistone, a polyhelical model of chromosome organization. *Proc. Natl. Acad. Sci. USA,* 1974 71(5), 1672-1676.

Tafvizi, A; Huang, F; Fersht, AR; Mirny, LA; van Oijen, AM. A single-molecule characterization of p53 search on DNA. *Proc. Natl. Acad. Sci. USA,* 2011 108(2), 563-568.

Talasz, H; Sarg, B; Lindner, H. Site-specifically phosphorylated forms of H1.5 and H1.2 localized at distinct regions of the nucleus are related to different processes during the cell cycle. *Chromosoma,* 2009 118(6), 693-709.

Talbert, PBL; Ahmad, K; Almouzni, G et al. A unified phylogeny-based nomenclature for histone variants. *Epigenetics Chromatin,* 2012 5, 7.

Tantos, A; Han, KH; Tompa, P. Intrinsic disorder in cell signaling and gene transcription. *Mol. Cell Endocrinol.,* 2012 348(2), 457-465.

Teufel, DP; Freund, SM; Bycroft, M; Fersht, AR. Four domains of p300 each bind tightly to a sequence spanning both transactivation subdomains of p53. *Proc. Natl. Acad. Sci. USA,* 2007 104, 7009–7014.

Thambirajah, AA; Ng, MK; Frehlick, LJ; Li, A; Serpa, JJ; Petrotchenko, EV; Silva-Moreno, B; Missiaen, KK; Borchers, CH; Adam-Hall, J; Mackie, R; Lutz, F; Gowen, BE; Hendzel, M; Georgel, PT; Ausió J. MeCP2 binds to nucleosome free (linker DNA) regions and to H3K9/H3K27 methylated nucleosomes in the brain. *Nucleic Acids Res.,* 2012 40(7), 2884-2897.

Th'ng, JP; Guo, XW; Swank, RA; Crissman, HA; Bradbury, EM; Inhibition of histone phosphorylation by staurosporine leads to chromosome decondensation. *J. Biol. Chem.*, 1994 269(13), 9568-9573.

Th'ng, JP; Sung, R; Ye, M; Hendzel, MJ. H1 family histones in the nucleus. Control of binding and localization by the C-terminal domain. *J. Biol. Chem.*, 2005 280(30), 27809-27814.

Uversky, VN. A decade and a half of protein intrinsic disorder, biology still waits for physics. *Protein Sci.*, 2013 22(6), 693-724.

Uversky, VN. Intrinsically disordered proteins and novel strategies for drug discovery. *Expert Opin Drug Discov.*, 2012 7(6), 475-488.

van Royen, ME; Cunha, SM; Brink, MC; Mattern, KA; Nigg, AL; Dubbink, HJ; Verschure, PJ; Trapman, J; Houtsmuller, AB. Compartmentalization of androgen receptor protein-protein interactions in living cells. *J. Cell Biol.*, 2007 177(1), 63-72.

Vila, R; Ponte, I; Collado, M; Arrondo, JL; Jiménez, MA; Rico, M; Suau P. DNA-induced alpha-helical structure in the NH2-terminal domain of histone H1 *J. Biol. Chem.*, 2001a 276(49), 46429-46435.

Vila, R; Ponte, I; Collado, M; Arrondo, JL; Suau P. Induction of secondary structure in a COOH-terminal peptide of histone H1 by interaction with the DNA: an infrared spectroscopy study. *J Biol Chem.* 2001b 276(33), 30898-30903

Vila, R; Ponte, I; Jiménez,MA; Rico, M; Suau, P. An inducible helix-Gly-Gly-helix motif in the N-terminal domain of histone H1e, a CD and NMR study. *Protein Sci.*, 2002 11(2), 214-220.

Vousden KH; Prives C. Blinded by the Light, The Growing Complexity of p53. *Cell,* 2009 137(3), 413-431.

Vuzman, D; Hoffman, Y; Levy, Y. Modulating protein-DNA interactions by post-translational modifications at disordered regions. *Pac. Symp. Biocomput.*, 2012, 188-199.

Vuzman, D; Levy, Y. Intrinsically disordered regions as affinity tuners in protein-DNA interactions. *Mol. Biosyst.*, 2012 8(1), 47-57.

Vyas, P; Brown, DT. N- and C-terminal domains determine differential nucleosomal binding geometry and affinity of linker histone isotypes H1(0) and H1c. *J. Biol. Chem.*, 2012, 287(15) 11778-11787.

Wakefield, RI; Smith, BO; Nan, X; Free, A; Soteriou, A; Uhrin, D; et al. The solution structure of the domain from MeCP2 that binds to methylated DNA. *J. Mol. Biol.*, 1999 291, 1055–1065.

Wang, L; Li, L; Zhang, H; Luo, X; Dai, J; Zhou, S; Gu, J; Zhu, J; Atadja, P; Lu, C; Li, E; Zhao, K. Structure of human SMYD2 protein reveals the basis of p53 tumor suppressor methylation. *J. Biol. Chem.*, 2011 286(44), 38725-38737.

Wang, X; Moore, SC; Laszckzak, M; Ausió, J. Acetylation increases the alpha-helical content of the histone tails of the nucleosome. *J. Biol. Chem.*, 2000 275(45), 35013-350120.

Wärnmark, A; Wikström, A; Wright, AP; Gustafsson, JA; Härd T. The N-terminal regions of estrogen receptor alpha and beta are unstructured in vitro and show different TBP binding properties. *J. Biol. Chem.*, 2001 276(49), 45939-45944.

Weinberg, RL; Freund, SM; Veprintsev, DB; Bycroft, M; Fersht, AR. Regulation of DNA binding of p53 by its C-terminal domain. *J. Mol. Biol.*, 2004a 342(3), 801-811.

Weinberg, RL; Veprintsev, DB; Fersht, AR. Cooperative binding of tetrameric p53 to DNA. *J. Mol. Biol.*, 2004b 341(5), 1145-1159.

Wright, PE; Dyson, HJ. Intrinsically disordered proteins in cellular signalling and regulation. *Nat. Rev. Mol. Cell Biol.*, 2015 16(1), 18-29.

Xie, H; Vucetic, S; Iakoucheva, LM; Oldfield, CJ; Dunker, AK; Obradovic, Z; Uversky VN. Functional anthology of intrinsic disorder. 3. Ligands, post-translational modifications, and diseases associated with intrinsically disordered proteins. *J. Proteome Res.*, 2007 6(5), 1917-1932.

Xu, YM; Du, JY; Lau, AT. Posttranslational modifications of human histone H3, an update. Proteomics, 2014 14(17-18), 2047-2060.

Yang, C; van der Woerd, MJ; Muthurajan,UM; Hansen, JC; Luger, K. Biophysical analysis and small-angle X-ray scattering-derived structures of MeCP2-nucleosome complexes. *Nucleic Acids Res.*, 2011 39(10), 4122-35.

Yang, N; Wang, W; Wang, Y; Wang, M; Zhao, Q; Rao, Z; Zhu, B; Xu, RM. Distinct mode of methylated lysine-4 of histone H3 recognition by tandem tudor-like domains of Spindlin1. *Proc. Natl. Acad. Sci. USA*, 2012 109(44), 17954-17959.

Yu, Q; Ye, W; Wang, W; Chen HF. Global conformational selection and local induced fit for the recognition between intrinsic disordered p53 and CBP. *PLoS One*, 2013 8(3)e59627.

Zhan, YA; Wu H; Powell, AT; Daughdrill, GW; Ytreberg, FM. Impact of the K24N mutation on the transactivation domain of p53 and its binding to murine double-minute clone 2. *Proteins, Structure, Function, and Bioinformatics*, 2013 81, 1738-1747.

Zhao Q; Chasse SA; Devarakonda S; Sierk ML; Ahvazi B; Rastinejad F. Structural basis of RXR-DNA interactions. *J. Mol. Biol.*, 2000 296(2), 509-520.

Zhao, Q; Khorasanizadeh, S; Miyoshi, Y; Lazar, M.A; Rastinejad, F. Structural elements of an orphan nuclear receptor–DNA complex. Mol. *Cell*, 1998 1, 849–861.

Zheng, Y; John, S; Pesavento, JJ; Schultz-Norton, JR; Schiltz, RL; Baek, S; Nardulli, AM; Hager, GL; Kelleher, NL; Mizzen, CA. Histone H1 phosphorylation is associated with transcription by RNA polymerases I and II. *J. Cell Biol.*, 2010 189(3), 407-415.

Zhou, XZ; Kops, O; Werner, A; Lu, PJ; Shen, M; Stoller, G; Küllertz, G; Stark, M; Fischer, G; Lu, KP. Pin1-dependent prolyl isomerization regulates dephosphorylation of Cdc25C and tau proteins. *Mol. Cell.*, 2000 6(4), 873–883.

Zlatanova, J; Caiafa, P; Van Holde, K. Linker histone binding and displacement, versatile mechanism for transcriptional regulation. *FASEB J.*, 2000 14(12), 1697-1704.

In: Intrinsically Disordered Proteins (IDPs)   ISBN: 978-1-63484-407-9
Editor: Violet Weber   © 2016 Nova Science Publishers, Inc.

Chapter 2

STRUCTURAL ANALYSIS OF INTRINSICALLY DISORDERED PROTEINS: COMPUTER ATOMISTIC SIMULATION

Anna Battisti[1], *Gabriele Ciasca*[2] *and Alexander Tenenbaum*[3,*]
[1]International School for Advanced Studies (SISSA), Trieste, Italy
[2]Physics Institute, Catholic University, Roma, Italy
[3]Physics Department, Sapienza University, Roma, Italy

Abstract

Intrinsically disordered proteins (IDPs) are biomolecules that do not have a definite 3D structure; their role in the biochemical network of a cell relates to their ability to switch rapidly among different secondary and tertiary structures. For this reason, applying a simulation computer program to their structural study turns out to be problematic, as their dynamical simulation cannot start from a known list of atomistic positions, as is the case for globular proteins that do crystallize and that one can analyze by X-ray spectroscopy to determine their structure.

We have established a method to perform a computer simulation of these proteins, apt to gather statistically significant data on their transient structures. The only required input to start the procedure is the primary sequence of the disordered domains of the protein, and the 3D structure of the ordered domains, if any. For a fully disordered protein the method is as follows:

---

*E-mail address: alexander.tenenbaum@roma1.infn.it

(a) The first step is the creation of a multi-rod-like configuration of the molecule, derived from its primary sequence. This structure evolves dynamically *in vacuo* or in an implicit model of solvent, until its gyration radius - or any other measure of the overall configuration of the molecule - reaches the experimental average value; at this point, one may follow two different paths.

(b1) If the study focuses on transient secondary structures of the molecule, one puts the structure obtained at the end of the first step in a box containing solvent molecules in explicit implementation, and a standard molecular dynamics simulation follows.

(b2) If the study focuses on the tertiary structure of the molecule, a larger sampling of the phase space is required, with the molecule moving in very large and diverse regions of the phase space. To this end, the structure of the IDP is let evolve dynamically in an implicit solvent using metadynamics, an algorithm that keeps track of the regions of the phase space already sampled, and forces the system to wander in further regions of the phase space.

(c) One can increase the accuracy of the statistical information gathered in both cases by fitting, where available, experimental data of the protein. In this step one extracts an ensemble of 'best' conformers from the pool of all configurations produced in the simulated dynamics. One derives this ensemble by means of an ensemble optimization method, implementing a genetic algorithm.

We have applied this procedure to the simulation of tau, one of the largest fully disordered proteins, which is involved in the development of Alzheimer's disease and of other neurodegenerative diseases. We have combined the results of our simulation with small-angle X-ray scattering experimental data to extract from the dynamics an optimized ensemble of most probable conformers of tau.

The method can be easily adapted to IDPs entailing ordered domains.

PACS: 87.14.E-, 87.15.ap, 87.15.bd, 87.15.bg

Keywords: Intrinsically Disordered Proteins; tau protein; transient secondary structures; transient tertiary structures; molecular dynamics; metadynamics

1. Introduction

Intrinsically disordered proteins (IDPs) do not have an average stable structure in their native state; they are similar to a random coil fluctuating in an ensemble of conformations, and resemble highly denatured proteins [1, 2].

Due to their flexibility, and at a variance with the well-known lock-and-key biomolecular paradigm, they perform tasks that globular proteins cannot perform [1, 3, 4, 5, 6]. IDPs entail at least one extended disordered region, and can entail globular domains alternating with flexible linkers or disordered domains. These proteins are therefore characterized by different degrees of disorder, from those formed by globular domains connected by disordered segments to those totally disordered [1, 2]. Even the latter may entail segments endowed, albeit temporarily, with secondary structures such as α-helices, β-sheets or PPII helices [7].

The main functions of IDPs are not structural, but regulatory: control, modulation and signalling. A characteristic feature of their biological function is an interaction energy among residues that is significantly lower than for globular proteins [1, 6]; this favors fast shifts between extended conformations, which generally accompany the binding to other molecules, and disordered molten globule-like conformations. The formation and dissolution of bound states is probably faster than in the case of globular proteins [1, 4].

The biochemical functions of IDPs relate to their secondary and tertiary structures, which vary in time. Given the speed at which transitions between different conformations are supposed to take place, a computer simulation of their dynamics seems to be a promising tool to characterize their time-dependent structure and to understand their behavior. The dynamical simulation of an IDP is a computational challenge, because by definition there are no experimentally determined 3D structures of the whole molecule, such as a Protein Data Bank file, from which to start. We have developed a method to perform computer simulations of IDPs, apt to overcome this obstacle and to gather statistically significant data on the transient structures of the proteins.

The simulation of an IDP confronts a second problem, namely the choice of a suitable force field. Molecular mechanics force fields have been parametrized on folded protein structures, and therefore may not correctly reproduce the structure of disordered proteins. On the other hand, there is no alternative to the use of one of the known force fields, because an *ab initio* calculation of a large disordered molecule would be unfeasible. Our simulations indicate the known force fields are apt to represent - with some caution - also the dynamics of IDPs.

The only required input to start the procedure is the primary sequence of the protein. We will illustrate the method by applying it to protein tau, a large, fully disordered protein[1].

2. The Tau Protein

The tau protein, one of the largest fully disordered IDPs [3], is involved in the nucleation and stabilization of the microtubules (MTs) in the cytoskeleton of the axons of the neurons. Protein tau achieves stabilization through the bonding of its repeats domain to the α- and β-tubulines forming the MTs [8]. But the same tau can aggregate in paired helical filaments (PHFs) and form fibrils which, in their turn, form insoluble tangles [1, 8]; this pathological deviation from its physiological function together with other factors triggers the development of Alzheimer's disease and of other neurodegenerative diseases [8].

Tau exists in several isoforms[1]; we have chosen to simulate its htau40 isoform, which is located in the human central nervous system; it has 441 residues and a molecular weight of 45.85 kDa. One can distinguish in its primary sequence four domains, corresponding to morphologically different sections of the molecule: the N-terminal projection domain (residues 1-150); a proline-rich segment (residues 151-243); a domain entailing four repeats (residues 244-368); the C-terminal domain (residues 369-441).

The process of formation of the PHFs is not entirely known, but some of its factors and stages have been investigated. A precursor stage of tau's polymerization has been related to specific transient global folds of the protein, in which the N-terminal is folded near the repeats domain in a hairpin conformation [7, 9, 10], or the C terminal is in proximity of the repeats domain and the N-terminal is folded near the C-terminal in a paperclip conformation [5, 11, 12]. The pathological aggregation in the form of insoluble tangles has been attributed to a local transition from the unfolded state to a β-structure [13, 14, 15, 16]. The aggregation process is supposed to start from a nucleus entailing the VQIVYK motif, a segment with high propensity for a β-structure [10], or the VQIINK motif. These two hexapeptides are, respectively, at the beginning of the third and of the second repeat [7, 9, 10], and have been identified as components of steric zippers formed by β-sheets parallel to the axis of the fibril [16]. The propensity of a polyproline-II motif toward the formation of β-sheets has also

---

[1] www.uniprot.org/uniprot/P10636; www.disprot.org/protein.php?id=DP00126

been listed among the possible causes of the aggregation of tau proteins, and hence of the origin of tau-pathologies [4, 17].

3. Initial Configuration and Dynamical Evolution

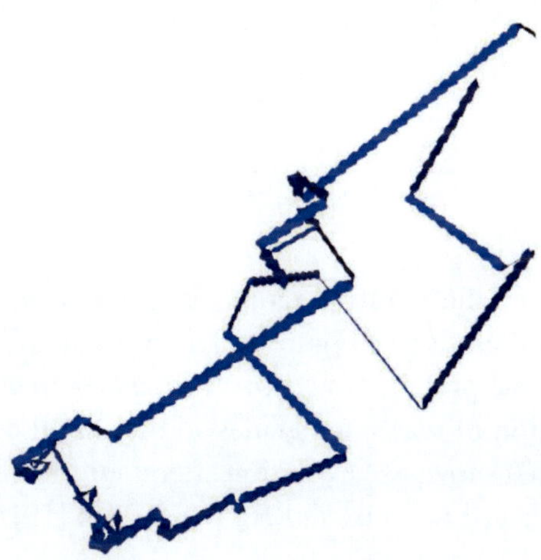

Figure 1. Initial shape of tau protein, as produced by the VMD program from the primary sequence [19].

In order to perform a molecular dynamics (MD) simulation of tau, we start from its primary sequence of amino acids. We use this sequence as an input to the visual MD (VMD) program [18]; this program lines up the amino acids following their primary sequence, departing from a straight line only when obliged by stereochemical incompatibility of neighboring amino acids. The output of VMD is thus a 3D sequence of straight segments of amino acids that bears little resemblance to a real protein, as shown in Fig.1. (There are other programs similar to VMD).

We use the MD simulation program GROMACS[2] to evolve this multi-rod-like structure *in vacuo* at $T = 300$ K [19]. The molecule's configuration collapses in a short time. This can be monitored by measuring the gyration radius $R_g = (\sum_i r_i^2 m_i / \sum_i m_i)^{1/2}$, where $\boldsymbol{r}_i$ are the positions of the atoms with

---

[2]GROMACS release 4.5.3, www.gromacs.org; box volume = 15253 nm3; ffamber99 force field; time step 2 fs; modified Berendsen thermostat, Parrinello-Rahman pressure coupling.

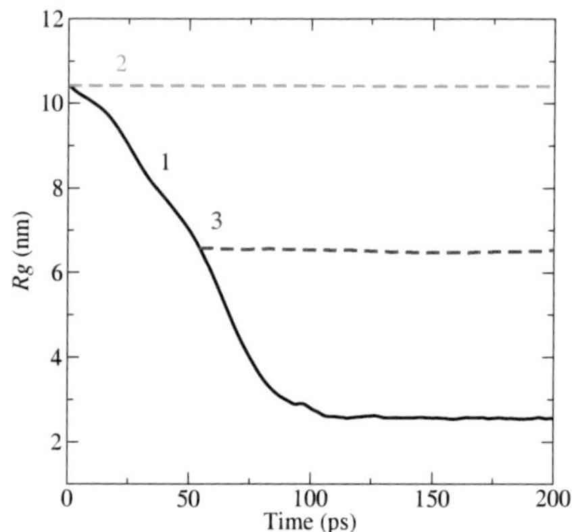

Figure 2. Evolution of the gyration radius at $T = 300\text{K}$ in the first 200 ps, starting from the configuration of Figure 1. The continuous black line (1) shows the rapid collapse of the protein *in vacuo*. The dashed green line (2) shows the evolution after addition of water molecules to the initial configuration of tau. The dashed red line (3) shows the evolution after addition of water molecules to the structure extracted at $t = 56$ ps and $R_g = 6.57$ nm [19].

respect to the center of mass of the molecule, and m_i are their masses; R_g measures the average size of the overall conformations of a molecule. Curve #1 in Figure 2 shows that the collapse of R_g, from an initial value of 10.4 nm to a value of 2.5 nm, takes place in about 100 ps and yields a very compact and entangled configuration. This fast evolution of the molecule is due to the absence of the solvent, which would prevent the collapse and the formation of a high number of intramolecular H-bonds as can be seen in Figure 3 (curve #1). We therefore stop the evolution when the configuration has reached a value of the gyration radius equal to the experimental average $R_g = 6.57$ nm [5]. The structure obtained in this way is then embedded in water, using an explicit or an implicit model.

The gyration radius is known for many molecules, being measured either by light scattering, or by SAXS (small-angle X-ray scattering), or by small angle neutron scattering. One could instead use any other measure of the overall configuration of the molecule to monitor its evolution *in vacuo*, like asphericity, which combines shape and compactness, and is measured by fluorescence microscopy [20].

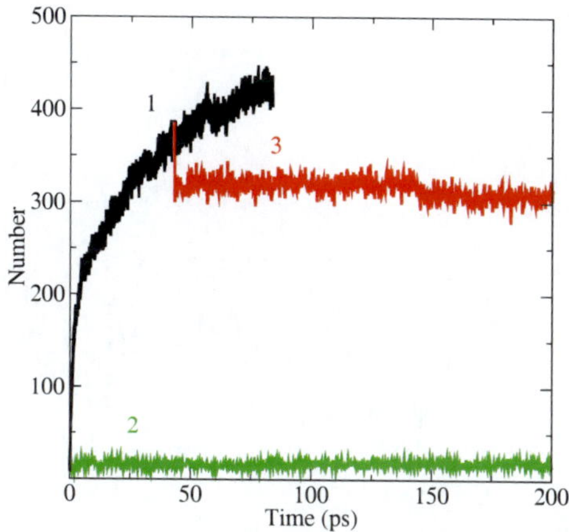

Figure 3. Time evolution of the number of intramolecular H-bonds at $T = 300$ K, first 200 ps. The black line (1) shows their rapid increase during the evolution *in vacuo*. The green line (2) shows their slow change after addition of water molecules to the initial configuration of tau. The red line (3) shows a sharp drop when water molecules are added to the structure extracted at t = 56 ps and the stable subsequent evolution [19].

An alternative way to reach in a short time a configuration of a large molecule from which to start the dynamical simulation is to embed the initial multi-rod-like molecule in an implicit solvent. When one immerses the initial VMD structure in implicit water [21], it rapidly evolves to a natural conformation, which can then be used to start a simulation. The time of this initial evolution is similar to the time of the evolution *in vacuo*; for tau, it takes about 50 ps.

A straightforward way of letting the initial multi-rod-like configuration evolve towards a native-like state of the molecule would be to embed its initial structure in explicit solvent. The configuration produced by VMD from the primary sequence is extended, and the simulation box has to be accordingly large, as a fully disordered protein (like tau) fluctuates in an ensemble of very different conformations. Therefore, in a MD simulation of such an IDP, one has to pay attention to the flexibility of the molecular structure: when periodic boundary conditions are used, the box must be large enough to avoid that during the dynamics the protein interacts with one of its periodic images, extending its

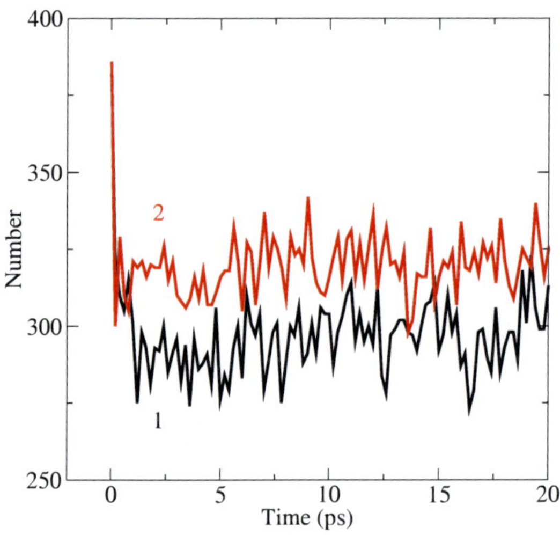

Figure 4. Time evolution of the number of intramolecular H-bonds at $T = 300$ K, beginning at the time when solvent is introduced in the simulation. Black line (1): explicit water molecules. Red line (2): implicit water solvent [19].

shape to the region bordering the walls of the box. A large box entails a very large number of solvent molecules; this is a relevant obstacle to make such a simulation workable. Using the box-to-molecule size relation usually adopted in this kind of MD simulation, in the case of tau one would have to use a box filled with about $1.5 \cdot 10^6$ water molecules.

Moreover, the evolution of the molecule from the initial configuration is very slow, as shown by curves #2 in Figure 2 and Figure 3. Taking into account the experimental value for the average gyration radius $R_g = 6.57$ nm, one can foresee for the overall configuration of the molecule, and for R_g, an equilibration time in excess of several tens of nanoseconds. All in all, this would be a computationally very expensive procedure for a molecule as large as tau; but it could be feasible for a small molecule.

4.   Secondary Structures

The initial evolution, either *in vacuo* or in implicit solvent, lasts few tens of ps; at the end one gets a structure of the molecule of a size comparable to the average experimental one. For large molecules (like tau) the size is significantly reduced with respect to the one produced by VMD; its structure can thus be put in a simulation box much smaller than the initial one, and the number of solvent molecules in the box is significantly reduced with respect to the number needed to embed the initial extended state produced via VMD.

At this stage the simulation run can begin, embedding the molecule's structure in a suitable solvent. Let us assume that the computer experiment aims at gathering information on the existence and probability of segments of the molecule endowed, albeit temporarily, with a secondary structure. As we show below, to achieve the best simulation of these structures the solvent must be represented by explicit molecules, in order to have a realistic competition between intramolecular and intermolecular (that is, between protein and water molecules) H-bonds.

Tau is a good candidate for this kind of simulation. Its value of R_g is large due to the molecule's non-globular structure, and is about that of a random coil of the same length (6.9 nm). Nevertheless, when R_g is measured in partial domains of tau that entail the repeats, its value turns out to be larger than the value estimated for a random coil; this hints at a propensity of these domains to form secondary structures [5, 11]. Because these structures would very likely be transient, there is a definite interest in a dynamical simulation of tau, in order to acquire a detailed knowledge of these transient patterns.

After a short minimization of the total energy, the system (tau + water molecules [3]) is ready to start a dynamical evolution in a region of the phase space corresponding to realistic conformations of the molecule. The stabilizing effect of the introduction of explicit water on the dynamical evolution of the protein implies a sudden interruption of the collapse of the molecule, and the beginning of a slow fluctuation of the structure; one can see this in Figure 2, curve #3. As for the intramolecular H-bonds, curve #3 of Figure 3 shows that the introduction of the water molecules causes a sudden decrease in the number of those bonds, about a fourth of which is replaced by H-bonds between tau and the water molecules. Figure 4 shows more in detail this instantaneous decrease (curve #1), and the first 20 ps of a NVT (constant number of molecules, volume, and temperature) simulation. After the introduction of the solvent, the molecule assumes in a short time a native-like configuration, as shown in Figure 5; the structure displays short transient secondary structures like β-sheets and

---

[3]spce water model in our simulation.

Figure 5. Shape of the tau protein after a 100 ps evolution, 56 ps *in vacuo* and 44 ps in explicit water, at $T = 300$ K. Two transient short β-sheets (yellow) and a transient short α-helix (purple) are highlighted [19].

α-helices.

One could follow a similar procedure by embedding in implicit water [21] the molecule's structure extracted midway during the collapse *in vacuo*, when the molecule has shrunk to a natural size, or leaving it in the implicit solvent, if one used this from the beginning. The use of an implicit solvent would allow a much faster simulation run, if compared to the implementation of an explicit water model. On the other hand, the two solvent types are known to operate differently in the prediction of secondary structures [22, 23]. Indeed, as shown in Figure 4, the effect of implicit water on the replacement of intramolecular H-bonds is not the same as that of explicit water molecules: the latter seem to be more efficient in competing with intramolecular H-bonds and replacing them with solvent-solute ones. After an evolution *in vacuo* of the molecule, possible spurious H-bonds, which would not form if a solvent had been included in the simulation from the beginning of the dynamics, are better removed by putting the molecule in explicit solvent.

We used the initial conformation of tau produced by VMD to start a simulation at constant temperature and pressure [24]. For this simulation we have chosen the ffG53a6 force field, implemented in the GROMACS package 3 [4]. The simulation has been carried out at neutral pH (pH = 7), close the physiological value (that is in the 7.2 - 7.4 range). Accordingly, amino acids were set to their default protonation states at pH = 7, with Lys, Arg carrying a +1 and Glu, Asp a -1 net charge.

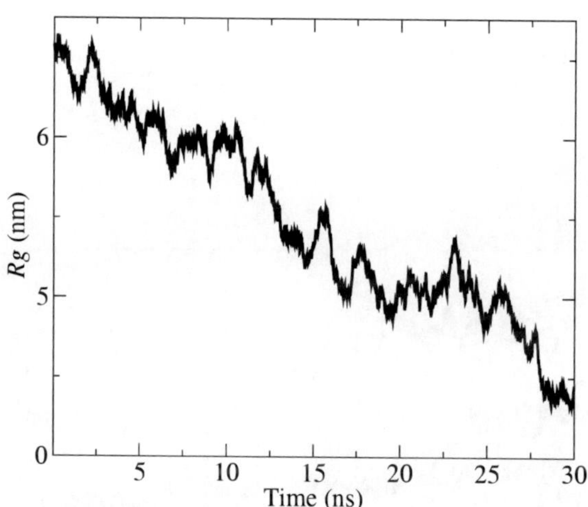

Figure 6. Time evolution of the gyration radius of protein tau during the standard MD dynamics at $T = 300$ K [24]. The experimental average value of R_g is 6.6 nm and the standard deviation is 0.3 nm [13].

In this first dynamical simulation of the complete tau (htau40), we have studied the time evolution of the molecule in water over a time of 30 ns. Figure 6 shows the gyration radius; R_g is not stable around its experimental value, as it progressively decreases to about 4.3 nm. Even though the latter value is within

[4] GROMACS release 4.5.3, www.gromacs.org; box volume = 15253 nm3; ffG53a6 force field; spce water model; time step 2 fs; modified Berendsen thermostat, Parrinello-Rahman pressure coupling.

the range of values computed from a set of static conformers of tau produced by the EOM method [5, 14], the continuous decrease of R_g hints at a possible short-coming of the force field in reproducing the overall shape of the molecule. In order to clarify this dynamical behavior we have computed the time evolution of the gyration radius of the four domains corresponding to morphologically differ-ent sections of the molecule: the N-terminal projection domain, the proline-rich segment, the repeats domain, and the C-terminal domain. We report the results in Figure 7; they show that all four domains reach an equilibrium stage: first the C-terminal domain, after about 10 ns; second the repeats domain, shortly before 20 ns; then the proline-rich segment and the N-terminal domain, after 22 ns. The final decrease of the total R_g visible in Figure 6 after an apparent stabilization between 18 and 24 ns has thus to be attributed to a reduction of the distances among domains, rather than to a further shrinking of one or more of them.

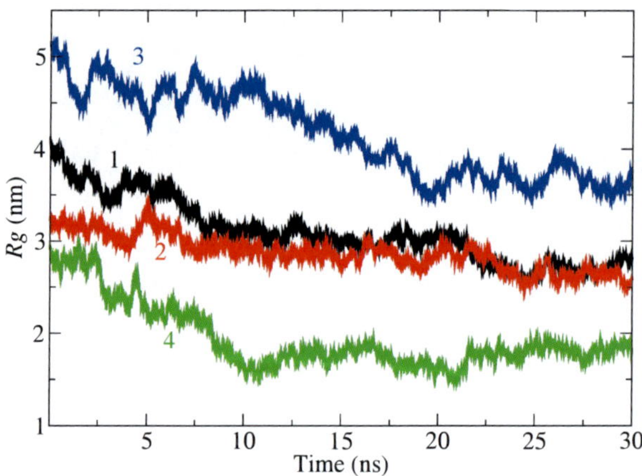

Figure 7. Time evolution of the gyration radius of four domains of tau at $T = 300$ K, in a standard MD simulation. Curve #1 (black): residues 1-150, N-terminal domain; curve #2 (red): residues 151-243, proline-rich segment; curve #3 (blue): residues 244-368, repeats domain; curve #4 (green): residues 369-441, C-terminal domain [24].

One can notice that the stabilized value of R_g found for the repeats domain in our simulation (3.7 nm) almost coincides with the experimental value of 3.8 nm found for the K18 construct of tau, the latter almost coinciding with the repeats domain [5].

In order to assess the propensities of various domains of tau to form temporary secondary structures, we have extracted information on the formation of temporary secondary structures from the whole 30 ns dynamics. We have measured the time evolution of the number of residues found in coils, β-sheets, β-bridges, bends, turns, and α-helices; we show these quantities are shown in Figure 8. While the majority of residues are in a coil-like conformation and in bends, there is a significant presence of secondary structures like turns, β-bridges, β-sheets, and α-helices.

The number of residues forming bends oscillates in a stable way during the dynamics; the number of residues forming turns stabilizes after about 6 ns; the number of residues forming a helix oscillates during the whole dynamics (mostly an α-helix, with some short shifts to a 3-helix or a 5-helix). Figure 8 shows that the formation of temporary secondary structures does not depend significantly on the overall shape of the molecule, due to their localized nature. The pattern of extension and time dependence of temporary secondary structures in tau shown in Figure 8 should thus be representative of the equilibrium state.

We have compared the average values of the extension and frequency of temporary α- and β-structures measured in this simulation with propensities to form α-helices or β-structures, assigned to various segments of tau using experimental NMR data [7]. Weighing the number of residues entailed in each of these segments with its propensity (fraction of time spent in the secondary structure), one finds an average number of 12 residues in β-structures and of 4 residues in an α-helix. These results are compatible with the results of our MD simulation, namely 26 ± 14 residues and 5 ± 2 residues for β-structures and α-helices, respectively [24].

5. Tertiary Structures

As mentioned before, particular tertiary structures, albeit transient, are supposed to play a key role in the pathological evolution of protein tau. These structures are likely to evolve over regions of the phase space that are much larger than those characteristic of temporary secondary structures. It is therefore important,

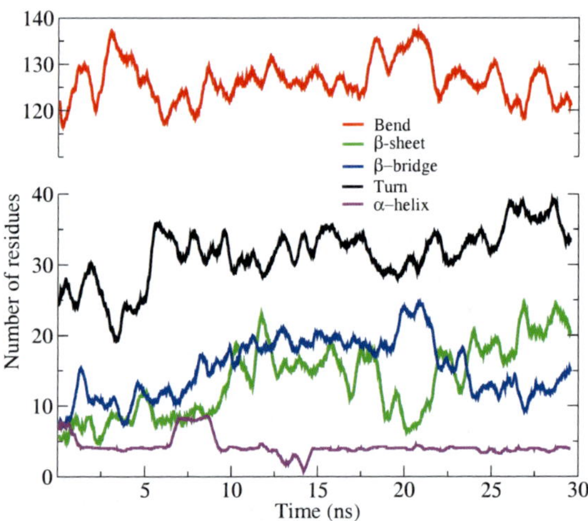

Figure 8. Number of residues found in secondary structures during the 30 ns dynamics. The curves have been smoothed by averaging the data over a sliding 1 ns interval [24].

in order to gather information on the existence and probability of configurations endowed with statistically significant tertiary structures, to achieve a dynamical simulation able to sample the overall structure of the molecule over large regions of the phase space.

The previous procedure, that has been proved to be effective in sampling local temporary secondary structures, would not be as effective in sampling larger regions of the space, as required to detect statistically relevant, albeit transient, tertiary structures. Metadynamics is a simulation tool that greatly expands the explored region of the phase space of an IDP, allowing a significant sampling of its fluctuating tertiary structure.

The metadynamics algorithm keeps track of the regions of the phase space already sampled by the molecule in its dynamics, recording a collective dynamical variable, and forces the system to leave those regions and to wander in other regions of the phase space [25]. The system thus samples a portion of the phase space much larger than the one sampled in an equal time of standard MD computation. Because of the heavier computation implied by keeping the dynamical track, the simulation of a large system, like protein tau, dictates the use of an

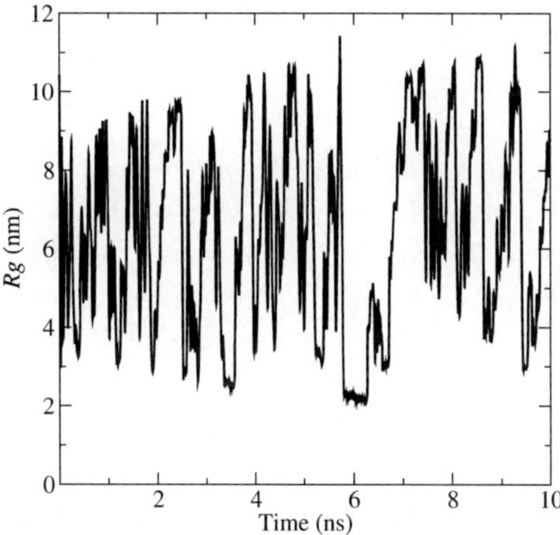

Figure 9. Time evolution of the gyration radius of protein tau during the metadynamics at $T = 300$ K [27].

implicit solvent model [26].

We have performed 10 ns of metadynamics simulation of protein tau [27]. The ffG53a6 force field we used in the previous molecular dynamics simulation of tau provides a statistical measure of local transient secondary structures; but the force field is less effective in maintaining over long times an extended conformation of the molecule. An advantage of the use of metadynamics is that it overcomes this problem, as its algorithm avoids the risk of a shrinking of the molecule by forcing its structure up and down the R_g range, as shown in Figure 9. We used in this simulation the ffamber99 force field, because the implicit solvent model could not be implemented with the ffG53a6 force field.

To implement this method we have chosen the collective dynamical variable R_g, the gyration radius. We have fixed the parameters of the metadynamics algorithm in such a way that R_g oscillates around its experimental value [5]. We have monitored the time evolution of tau by computing R_g; Figure 9 shows the gyration radius during a 10 ns evolution. The metadynamics algorithm induces large structural changes in the molecule, with the gyration radius spanning the

---

[5] CV = R_g; deposition stride $\tau = 10$ ps; height $W = 0.5$ kJ/mol; Gaussian width $\sigma = 0.35$ nm; limits on R_g: upper UWALL = 7.0 nm, lower LWALL = 5.5 nm.

range between 2.5 nm and 11 nm, centered near the experimental value: its average value over this evolution is $R_g = 6.3$ nm. The large oscillations of R_g hint at the variety of configurations sampled by the system in different regions of the phase space.

One can better understand the way in which the metadynamics algorithm acts on the system by separately computing the time evolution of the gyration radius of the N-terminal domain, of the proline-rich segment, of the repeats domain, and of the C-terminal domain. We report the results in Figure 10, which shows that all four domains undergo very strong modifications. The algorithm shakes them alternatively, leaving from time to time one of the domains in what appears to be a local equilibrium well. As an example, the proline-rich domain is almost stable during the first 2 ns, while the other domains show strong changes in their overall configuration; the first domain reaches again a relative stability between 3.5 and 6.3 ns, and between 9.0 and 9.6 ns. The other domains also stay for some time in a quasi-equilibrium state: the C-terminal between 5.2 and 6.3 ns, the N-terminal between 5.8 and 6.8 ns, and the repeats domain between 5.2 and 6.4 ns. The large amplitude of the oscillations of the N-terminal are due to its higher flexibility in comparison to the other domains, in particular the domain entailing the repeats [7].

The metadynamics algorithm drives the system to distant points of the space phase, improving the statistical sampling by producing likely and less likely configurations. Due to this drive, transient tertiary structures last short times, probably shorter than their natural lifetime in an unbiased dynamics. Therefore, a contact map of the molecule averaged over all configurations produced by the metadynamics simulation could highlight only resilient tertiary structures. A contact map of tau computed during the 10 ns metadynamics trajectory clearly shows a statistically relevant tertiary pattern entailing a long segment encompassing the N-terminal and the proline-rich domain (residues 120-190) in anti-parallel proximity of a segment encompassing the proline-rich domain and the repeats domain (residues 200-280) [27]. There must be a turn joining these two segments around residue 195, right in the middle of the proline-rich domain. Experiments have shown that the approach of the N-terminal to the central region of the molecule is involved in the aggregation process leading to the formation of PHFs [29]. It is also noteworthy that the end of this hairpin structure (residues 275-280) is the hexamer VQIINK, also known to be involved in this aggregation process [13, 14, 15].

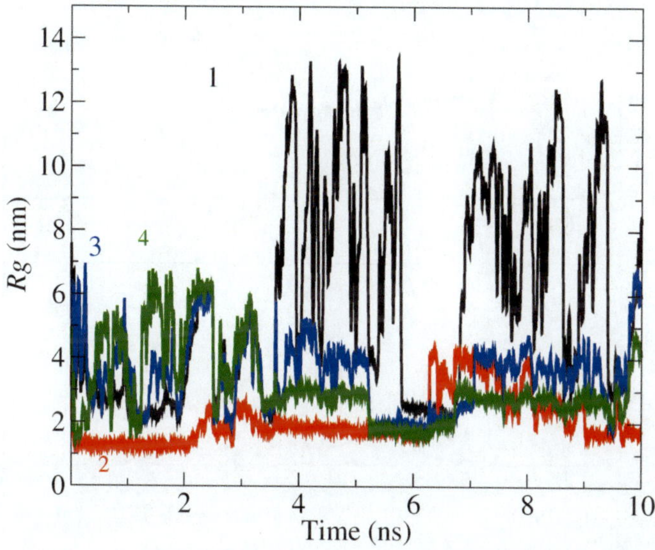

Figure 10. Time evolution of the gyration radius of four domains of tau at $T = 300$ K, in a metadynamics simulation. Curve labels and colors as in Figure 7 [27].

6. Fit of Experimental SAXS Data

We have improved the information on the equilibrium behavior of tau, obtained through the data produced by the computer simulation, by comparing them with experimental SAXS results obtained from a specimen of tau in solution. The SAXS experiment has been performed using full-length htau40. SAXS measurements were acquired on the BioSAXS beamline (ID 14-3) at the Synchrotron Radiation Facility ESRF (Grenoble, France) [30], at the constant temperature of 303 K. Solvent scattering was measured to allow an accurate subtraction of the background scattering. Figure 11(a) shows the result of this experiment. More details on the experiment are given in [28].

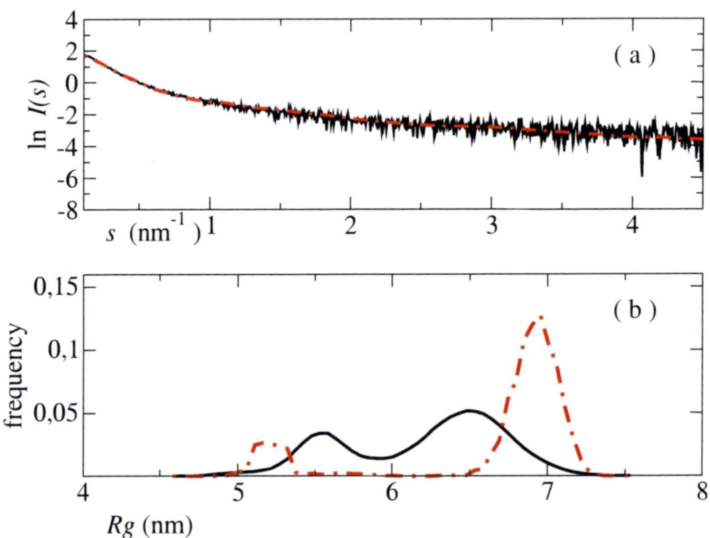

Figure 11. Panel (a): experimental SAXS curve (black continuous line); fit by an ensemble of conformers produced in a 30 ns standard MD simulation and selected by the genetic algorithm GAJOE (red dashed line). Panel (b): distribution of R_g values. Conformers produced by the simulation (black continuous line) and conformers selected by the genetic algorithm (red line), after addition of the coordinated water layer [24].

NMR could provide an alternative set of experimental data to compare with the results of the computer simulation. The program PALES, an atomic resolution approach to alignment tensor prediction, allows the computation of alignment tensors for a given configuration of the molecule; the same program calculates residual dipolar couplings that can be compared to the experimental ones [14].

We have used the SAXS curve to extract from our standard MD simulation [24] and from the metadynamics simulation [27] an ensemble of conformers that give the best fit of the experimental data.

The fitting procedure is as follows: (i) we extract about 9000 regularly spaced conformers of tau from our 30 ns standard MD simulation, or 10000 conformers of tau, regularly spaced by a 1 ps interval, from our 10 ns metadynamics simulation; (ii) this pool of conformers is processed by the program CRYSOL [31] to obtain the theoretical SAXS pattern of each conformer, taking properly into account the scattering from the hydration shell of the water layer coordinated with the molecule; (iii) the ensemble optimization method EOM with the genetic algorithm GAJOE [32] is then employed to select from the pool of theoretical SAXS curves an ensemble of conformers (162 from the standard MD run, 194 from the metadynamics run) that provide with their averaged theoretical scattering intensity the best fit of the experimental SAXS data; each conformer is weighed with its genetic multiplicity [32, 33].

The application of the genetic algorithm GAJOE over 1000 cycles to select the best ensemble of conformers progressively yields an improved fit of the SAXS data [6]. The final values of χ^2 attest the accuracy of the fit: 1.4 for the standard MD run, 1.0 for the metadynamics run. The statistical sampling of tau's phase space achieved by metadynamics is more accurate than the one achieved by standard molecular dynamics. This is an expected result, due to the larger portion of the phase space sampled by the metadynamics simulation.

As shown in Figure 11(a) and Figure 12(a), the selected ensembles fit quite well the experimental SAXS results, when each conformer is weighed with the appropriate multiplicity determined by the genetic algorithm. We show in Figure 11(b) and Figure 12(b) the distributions of R_g values of both the original pool (black line) and of the selected ensemble (red line), respectively. It may be noted that most selected conformers in the standard MD run belong to the first 5 ns of the trajectory, where the value of R_g is near to its initial equilibrium value; but there is also a significant presence of conformers with R_g values between 5.1 and 5.4 nm, belonging to the temporarily stabilized trajectory stretch between 18 and 24 ns (Figure 6). In the metadynamics run the R_g values of the ensemble selected by the genetic algorithm show a distribution peaked near the experimental value $R_g = 6.6$ nm, with a significant presence of conformers with R_g values between 3.0 and 10.5 nm. The radius of gyration averaged over this ensemble is $R_g = 6.8$ nm, with a standard deviation of 1.7 nm. It is in very good agreement with the theoretical value expected for a 441 amino acids random coil in solution, which is 6.9 nm [34].

---

[6] GAJOE parameters were set as follows: number of generations 1000; number of ensembles 50; number of curves per ensemble 20; number of mutations per ensemble 10; number of crossings per generation 20.

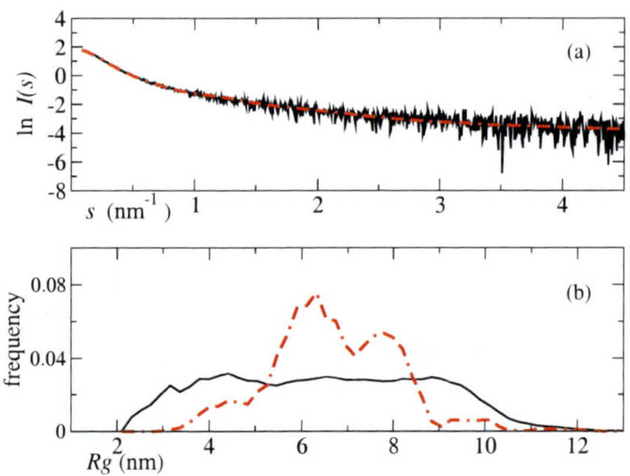

Figure 12. As in Figure 11, but for an ensemble of conformers produced in a 10 ns metadynamics simulation [27].

The distribution of the R_g values before and after the selection performed by means of EOM with the genetic algorithm reflects the higher efficiency of metadynamics in sampling different regions of the phase space, in comparison to a standard MD simulation of similar duration. This can be seen comparing panels in Figure 11(b), standard MD, and in Figure 12(b), metadynamics. The original pool in panel 12(b) encompasses a much broader range of R_g values than in panel 11(b). The selected ensemble distribution shown in Figure 12(b) appears to be broader and therefore statistically more significant than the one shown in Figure 11(b), and it is also approximately centered on the experimental value $R_g = 6.6$ nm.

7.   Transient Structures

The ensemble of 162 conformers selected from the standard MD run has been analyzed with the DSSP program [35, 36], as implemented in GROMACS, to identify secondary structures like coils, β-sheets, β-bridges, bends, turns, and α-helices. The propensity of the molecule to form these secondary structures, measured by the number of residues in each structure, turns out to coincide with the average number measured during the whole 30 ns MD run, within one standard deviation of the latter. This confirms the validity of the dynamical simulation as far as local secondary structures are concerned, notwithstanding a possible shortcoming of the force field with regard to the overall shape.

Comparing again our results with those obtained by Mukrasch and coworkers [7], we find a better agreement than in the standard MD run: 15 residues in β-structures and 6 residues in α-helices, to compare with the experimental results: 12 residues and 4 residues, respectively.

Figure 9 and Figure 10 clearly show the thorough shaking of the molecular structure produced by the metadynamics algorithm. This dynamics drives the system to distant points of the phase space and thus to very different global folds. Figure 13 displays the instant contact maps ($C_\alpha - C_\alpha$ distance between all pairs of residues) of four configurations, chosen among the pool of 194 conformers selected by the genetic algorithm. Three are among those selected with highest frequency (panels (a), (c), and (d)); the fourth (panel (b)) further illustrates the variety of global folds. Only distances smaller than 1.5 nm are on display.

The hairpin pattern described before is embedded in various hairpin or paperclip transient tertiary structures, as shown in panels (b), (c), and (d) of Figure 13. While all those configurations are transient, they share this common, persistent tertiary motif: a hairpin folding encompassing part of the N terminal, the proline-rich domain, the first repeat, and a functionally relevant part of the second repeat. As mentioned before, hairpin configurations [7, 9, 10] and paperclip configurations [5, 11, 12] are believed to be a precursor stage of the polymerization of tau leading to the formation of fibrils and to the onset of neurodegenerative diseases.

8.   Conclusion

Given the speed at which transitions between conformations of an IDP are supposed to take place, the computer simulation of their dynamics seems to be a promising tool to understand their behavior. We present a method that can be used for any IDP. In order to start a MD simulation of a fully disordered protein of unknown 3D structure, one begins from its primary sequence of amino acids. One then implements the following procedure to produce a 3D structure to start the simulation. (i) One creates a first 3D structure by feeding the VMD program (or a similar program) with the primary sequence of the whole protein. (ii) The resulting structure - a multi-rod-like sequence of amino acids

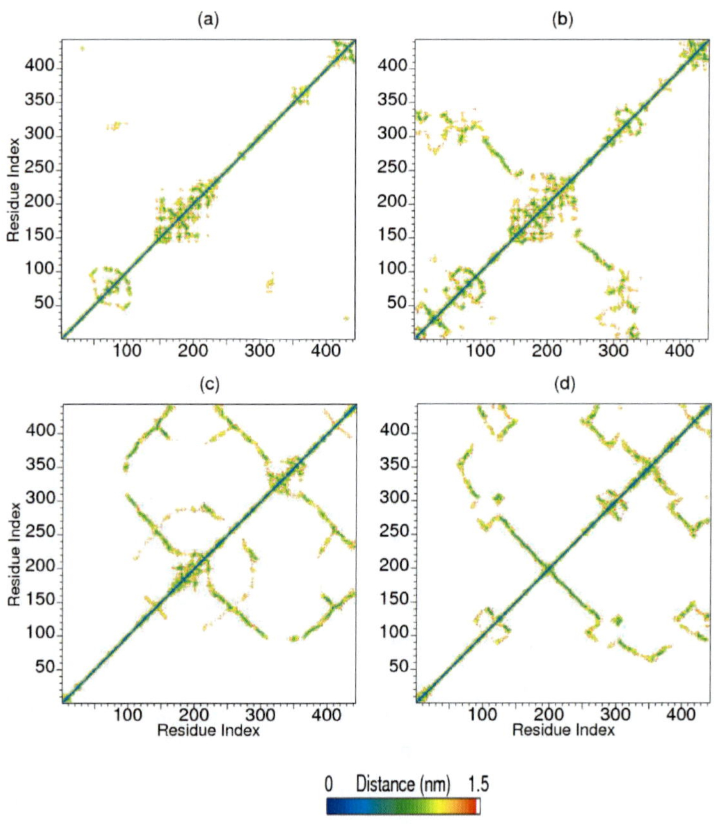

Figure 13. Instant contact maps ($C_\alpha - C_\alpha$ distance between pairs of residues) at $t = 41$ ps (a), $t = 1027$ ps (b), $t = 4228$ ps (c), $t = 7567$ ps (d).

- is put in a large box with periodic boundary conditions; after a short energy minimization this structure is taken as the initial one for a dynamical evolution *in vacuo* or in implicit solvent at the chosen temperature, performed with the package GROMACS or with any other simulation program; this step produces a rapid contraction of the protein. (iii) The evolution is stopped when the decreasing gyration radius R_g (or a similar shape-dependent variable) has reached its average experimental value. This yields a starting point for the simulation in a more realistic environment, i.e. with the addition of solvent. (iv) The simulation box is reduced to fit the reduced size of the protein and filled with solvent, either explicit or implicit. This significant reduction of the volume of the box greatly reduces the number of solvent molecules needed to fill it in the case of explicit solvent. (v) The energy of the system (protein + solvent) is then minimized in the case of explicit solvent, to allow the solvent molecules to adapt to the shape of the solute molecule. (vi) A short equilibration (about 100 ps) is performed at constant temperature. (vii) Another short equilibration (about 100 ps) is performed at constant temperature and pressure. (viii) One uses the last conformation of the previous step to start an extended simulation at constant temperature and pressure.

If the simulation focuses on the analysis of secondary structures, the best simulation is a standard MD one in explicit solvent. If the simulation focuses on the analysis of tertiary structures, the best simulation is metadynamics in implicit solvent.

The statistical information gathered in both cases can be refined by fitting, where available, experimental data of the protein, like small-angle X-ray scattering results. In this step one extracts an ensemble of 'best' configurations from the pool of all configurations produced in the simulated dynamics. This ensemble is produced by means of an ensemble optimization method, implementing a genetic algorithm. This set of conformers, selected from the simulated dynamics by fitting the experimental data, is the best approximation of an equilibrium ensemble that can be extracted from the simulated dynamics; it provides a significant observation of secondary structures and of the overall fold of the molecule. Furthermore, the selected ensemble of conformers represents a 3D data basis that one can use to start further simulations.

When applied to protein tau this method shows that the protein samples a limited number of almost stable secondary structure motifs (mainly short α and β structures), whereas the overall preferred conformation is a random coil entailing a hairpin motif. The transient nature of these structures in protein tau is relevant, and has biochemical implications *in vivo*: the hairpin motif is supposed to be involved in the early stages of tau's polymerization, leading to the formation of pathogenic paired helical filaments.

If the protein under study is only partially disordered, entailing regions of stable and known 3D structure, the method can be implemented by modifying only step (i): the program VMD is used to create multi-rod-like 3D structures of the disordered regions, which are then connected to the ordered regions. The following steps are unchanged,

Acknowledgment

This chapter derives in part from three articles published in Molecular Simulation on 1/2/2012, 26/4/2012, and 24/10/2013, copyright Taylor & Francis, available online at:
http://www.tandfonline.com/10.1080/08927022.2011.608671
http://www.tandfonline.com/10.1080/08927022.2011.633347
http://www.tandfonline.com/10.1080/08927022.2013.794275

References

[1] Tompa, P. Intrinsically disordered proteins. In: Sussman J, Silman I editors. *Structural Proteomics and its Impact on the Life Sciences.* Singapore: World Scientific; 2008; pp. 153-180.

[2] Dunker AK, Brown CJ, Lawson JD, Iakoucheva LM, and Obradovic Z. Intrinsic Disorder and Protein Function. *Biochemistry*, 2002, 41, pp. 6573-6582.

[3] Sickmeier M, Hamilton JA, LeGall T, Vacic V, Cortese MS, Tantos A, Szabo B, Tompa P, Chen J, Uversky VN, Obradovic Z, and Dunker AK. DisProt: the Database of Disordered Proteins. *Nucl. Acids Res.*, 2007, 35, pp. D786-D793.

[4] Tompa P. Intrinsically Unstructured Proteins. *Trends Biochem. Sci.*, 2002, 27, pp. 527-533.

[5] Mylonas E, Hascher A, Bernadó P, Blackledge M, Mandelkow E, and Svergun DI. Domain Conformation of Tau Protein Studied by Solution Small-Angle X-ray Scattering. *Biochemistry*, 2008, 47, pp. 10345-10353.

[6] Tompa P. The interplay between structure and function in intrinsically unstructured proteins. *FEBS Lett.*, 2005, 579, pp. 3346-3354.

[7] Mukrasch MD, Bibow S, Korukottu J, Jeganathan S, Biernat J, Griesinger C, Mandelkow E, and Zweckstetter M. Structural Polymorphism of 441-Residue Tau at Single Residue Resolution. *PLoS Biol.*, 2009, 7, pp. 399-414.

[8] Avila J, Lucas JJ, Pérez M, and Hernández F. Role of Tau Protein in Both Physiological and Pathological Conditions. *Physiol. Rev.*, 2004, 84, pp. 361-384.

[9] Carmel G, Mager EM, Binder LI, and Kuret J. The structural basis of monoclonal antibody Alz50s selectivity for Alzheimers disease pathology. *J. Biol. Chem.*, 1996, 271, pp. 32789-32795.

[10] Gamblin TC, Berry RW, and Binder LI. Tau Polymerization: Role of the Amino Terminus. *Biochemistry*, 2003, 42, pp. 2252-2257.

[11] Jeganathan S, von Bergen M, Brutlach H, Steinhoff HJ, and Mandelkow E. Global Hairpin Folding of Tau in Solution. *Biochemistry* 2006, 45, pp. 2283-2293.

[12] Bibow S, Mukrasch MD, Chinnathambi S, Biernat J, Griesinger C, Mandelkow E, and Zweckstetter M. The dynamic structure of filamentous tau. *Angew. Chem. Int. Ed.*, 2011, 50, pp. 11520-11524.

[13] von Bergen M, Barghorn S, Biernat J, Mandelkow E-M, and Mandelkow E. Tau aggregation is driven by a transition from random coil to beta sheet structure. *Biochimica et Biophysica Acta*, 2005, 1739, pp. 158-166.

[14] Mukrasch MD, Markwick P, Biernat J, von Bergen M, Bernadó P, Griesinger C, Mandelkow E, Zweckstetter M, and Blackledge M. Highly Populated Turn Conformations in Natively Unfolded Tau Protein Identified from Residual Dipolar Couplings and Molecular Simulation. *J. Am. Chem. Soc.*, 2007, 129, pp. 5235-5243.

[15] von Bergen M, Friedhoff P, Biernat J, Heberle J, Mandelkow E-M, and Mandelkow E. Assembly of tau protein into Alzheimer paired helical filaments depends on a local sequence motif ((306)VQIVYK(311)) forming beta structure. *Proc. Natl. Acad. Sci. USA*, 2000, 97, pp. 5129-5134.

[16] Sawaya MR, Sambashivan S, Nelson R, Ivanova MI, Sievers SA, Apostol MI, Thompson MJ, Balbirnie M, Wiltzius JJW, McFarlane HT, Madsen AØ, Riekel C, and Eisenberg D. Atomic structures of amyloid cross-spines reveal varied steric zippers. *Nature*, 2007, 447, pp. 453-457.

[17] Friedhoff P, von Bergen M, Mandelkow E-M, and Mandelkow E. Structure of tau protein and assembly into paired helical filaments. *Biochimica et Biophysica Acta*, 2000, 1502, pp. 122-132.

[18] Humphrey W, Dalka A, and Schulten K. Visual Molecular Dynamics. *J. Mol. Graph.*, 1996, 14, pp. 33-38.

[19] Battisti A and Tenenbaum A. Molecular dynamics simulation of intrinsically disordered proteins. *Mol. Simulat.*, 2012, 38, pp. 139-143.

[20] Arteca GA, Reimann CT, and Tapia O. Proteins *in vacuo*: Denaturing and folding mechanisms studied with computer simulated molecular dynamics. *Mass Spectrosc. Rev.*, 2001, 20, pp. 402-422.

[21] Still C, Tempczyk A, Hawley RC, and Hendrickson T. Semianalytical Treatment of Solvation for Molecular Mechanics and Dynamics. *J. Am. Chem. Soc.*, 1990, 112, pp. 6127-6129.

[22] Roe DR, Okur A, Wickstrom L, Hornak V, and Simmerling C. Secondary Structure Bias in Generalized Born Solvent Models: Comparison of Conformational Ensembles and Free Energy of Solvent Polarization from Explicit and Implicit Solvation. *J. Phys. Chem. B*, 2007, 111, pp. 1846-1857.

[23] Tan C, Yang L, and Luo R. How Well Does Poisson-Boltzmann Implicit Solvent Agree with Explicit Solvent? A Quantitative Analysis. *J. Phys. Chem. B*, 2006, 110, pp. 18680-18687.

[24] Battisti A, Ciasca G, Grottesi A, Bianconi A, and Tenenbaum A. Temporary secondary structures in tau, an intrinsically disordered protein. *Mol. Simulat.*, 2012, 38, pp. 525-533.

[25] Laio A and Gervasio FL. Metadynamics: a method to simulate rare events and reconstruct the free energy in biophysics, chemistry and material science. *Rep. Prog.Phys.*, 2008, 71, 126601 (22 pp).

[26] Still WC, Tempczyk A, Hawley RC, and Hendrickson T. Semianalytical treatment of solvation for molecular mechanics and dynamics. J. Am. Chem. Soc., 1990, 112, pp. 6127-6129.

[27] Battisti A, Ciasca G, and Tenenbaum A. Transient tertiary structures in tau, an intrinsically disordered protein. *Mol. Simulat.* 2013, 39, pp. 1084-1092.

[28] Ciasca G, Campi G, Battisti A, Rea G, Rodio M, Papi M, Pernot P, Tenenbaum A, and Bianconi A. Continuous thermal collapse of the intrinsically disordered protein tau is driven by its entropic flexible domain. *Langmuir*, 2012, 28, pp. 13405-13410.

[29] Gamblin TC, Berry RW, and Binder LI. Tau Polymerization: Role of the Amino Terminus. *Biochemistry* 2003, 42, pp. 2252-2257.

[30] Pernot P, Theveneau P, Giraud T, Nogueira RF, Nurizzo D, Spruce D, Surr J, McSweeney S, Round A, Felisaz F, Foedinger L, Gobbo A, Huet J, Villard C, and Cipriani F. New beamline dedicated to solution scattering from biological macromolecules at the ESRF. *J. Phys.: Conf. Ser.*, 2010, 247, 012009 (8 pp).

[31] Svergun DI, Barberato C, and Koch MHJ. CRYSOL a program to evaluate X-ray solution scattering of biological macromolecules from atomic coordinates. *J. Appl. Crystallogr.*, 1995, 28, pp. 768-773.

[32] Bernadó P, Mylonas E, Petoukhov MV, Blackledge M, and Svergun DI. Structural Characterization of Flexible Proteins Using Small-Angle X-ray Scattering. *J. Am. Chem. Soc.*, 2007, 129, pp. 5656-5664.

[33] EOM manual http://www.embl-hamburg.de/biosaxs/eom.html

[34] Kohn JE, Millett IS, Jacobs J, Zagrovic B, Dillon TM, Cingel N, Dothager RS, Seifert S, Thiyagarajan P, Sosnick TS, Hasan MZ, Pande VS, Ruczinski I, Doniach S, and Plaxco KW. Random-coil behavior and the dimensions of chemically unfolded proteins. *Proc. Natl. Acad. Sci. USA*, 2004, 101, pp. 12491-12496.

[35] Kabsch W and Sander C. Dictionary of protein secondary structure: pattern recognition of hydrogen-bonded and geometrical features. *Biopolymers*, 1983, 22, pp. 2577-2637.

[36] Joosten RP, te Beek TAH, Krieger E, Hekkelman ML, Hooft RWW, Schneider R, Sander C, and Vriend G. A series of PDB related databases for everyday needs. *Nucleic Acids Res.*, 2011, 39, pp. D411-D419.

In: Intrinsically Disordered Proteins (IDPs)   ISBN: 978-1-63484-407-9
Editor: Violet Weber   © 2016 Nova Science Publishers, Inc.

Chapter 3

BIOPHYSICAL CHARACTERIZATION OF INTRINSICALLY DISORDERED PROTEINS AND THEIR COMPLEXES: CHALLENGES AND CONTRADICTIONS

Alexander B. Sigalov[*]
SignaBlok, Inc., Shrewsbury, MA, US

ABSTRACT

The emergence of intrinsically disordered proteins (IDPs), the proteins that do not adopt a well-defined three-dimensional structure under physiological conditions, has challenged the classical protein structure-function paradigm. IDPs play an important role in cellular regulation, signaling and control in health and disease. However, the unusual biophysics of these proteins makes structural characterization of IDPs and their complexes not only challenging but often resulting in opposite conclusions. Here, I focus on two long-standing contradictions in the literature concerning dimerization and membrane-binding activities of IDPs. Molecular explanation of these discrepancies is provided. I also demonstrate how resolution of these critical issues in the field results in our expanded understanding of cell function with multiple applications in biology and medicine.

---

[*] Corresponding author: Alexander B. Sigalov. SignaBlok, Inc., Shrewsbury, MA 01545, US. E-mail address: sigalov@signablok.com.

INTRODUCTION

In the past 20 or so years, the central dogma of modern structural biology has assumed that three-dimensional protein structure determines its function. This notion has been challenged by the emergence of intrinsically disordered proteins (IDPs), which are proteins that lack a well-defined three-dimensional structure under physiological conditions (Dunker et al. 2002). Further studies of IDPs triggered the development of a new paradigm, the coupled binding and folding mechanism, that assumes IDPs adopt folded structures upon binding to their targets in order to perform their diverse biological functions (Dyson and Wright 2002). Coupled binding and folding can involve just a few residues or an entire protein domain. This subject has been addressed in detail in many comprehensive reviews and other articles (Dunker et al. 2008, Sigalov 2011, Tompa 2005, Uversky and Dunker 2010).

Intriguingly, structural studies of IDPs revealed unusual biophysical phenomena that substantially complicated these findings.

Here, I will focus on a new family of IDPs, cytoplasmic domains of immune receptors and cytoplasmic regions of signaling subunits from immune receptors, including those of ζ and CD3ε signaling subunits (ζ_{cyt} and CD3ε_{cyt}, respectively) of T cell receptor (TCR), and γ signaling subunit of FcεRI receptor (FcεRIγ_{cyt}) (Aivazian and Stern 2000, Sigalov et al. 2004) and discuss two long-standing contradictions and discrepancies in the literature concerning structural features and biophysical activities of these IDPs.

DO IDPS FOLD UPON BINDING TO THE CELL MEMBRANE?

Immune signaling-related cytoplasmic intrinsically disordered regions (IDRs) all have one or more copies of an immunoreceptor tyrosine-based activation motif (ITAM), tyrosine residues of which are phosphorylated upon receptor engagement in an early and obligatory event in the signaling cascade. Considering the crucial role of ζ_{cyt}, CD3ε_{cyt}, and FcεRIγ_{cyt} in immune signaling and their close proximity to the cell membrane, the question whether or not membrane binding of these IDPs can promote folding of ζ_{cyt}, CD3ε_{cyt}, and FcεRIγ_{cyt} ITAMs and thus lead to inaccessibility of the ITAM tyrosines for phosphorylation is of fundamental importance in our understanding of receptor triggering.

However, little is known about the lipid binding activity of ITAM-containing cytoplasmic domains and the existing data are strikingly contradictory (Figure 1).

LIPID BINDING OF IDPS

INAPPROPRIATE MODELS   **APPROPRIATE MODELS**

A. Different mechanisms of binding
(shown for ζ_{cyt}; similar for CD3ε_{cyt} and FcεRIγ_{cyt})

Coupled binding and folding   **Binding without folding**

Micelles   Unstable lipid bilayers   Stable lipid bilayers

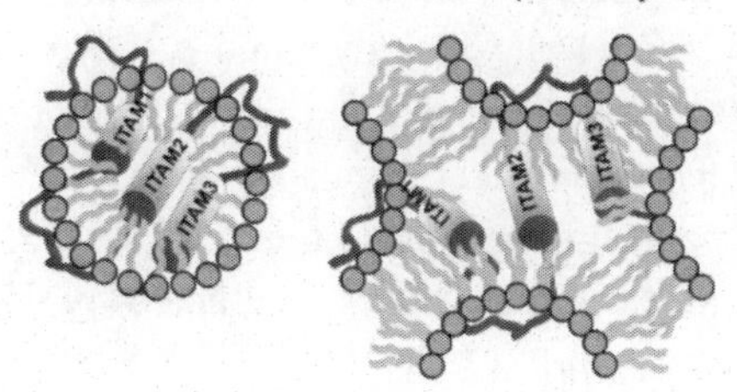

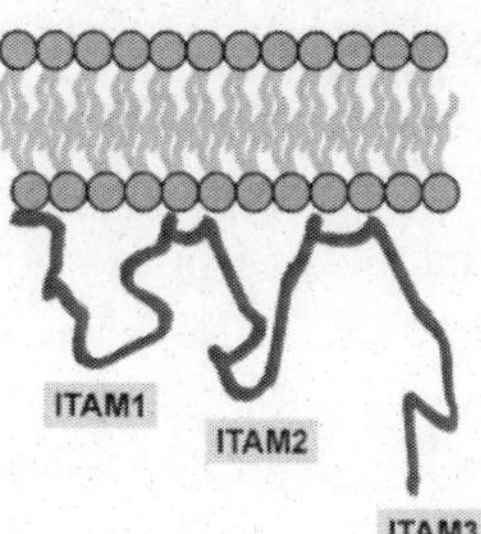

B. Contradictory conclusions

ITAMs are involved in binding	ITAMs are <u>not</u> involved in binding
ITAMs fold into helixes upon binding	No folding of ITAMs upon binding
Folding of ITAMs plays a role *in vivo*	No folding of ITAMs *in vivo*

ζ_{cyt} (Aivazian and Stern, 2000)
CD3ε_{cyt} (Xu et al., 2008)
CD3ε_{cyt} (Shi et al., 2013)

ζ_{cyt}, CD3ε_{cyt}, and FcεRIγ_{cyt}
(Sigalov et al., 2006, 2009)
ζ_{cyt} (DeFord-Watts et al., 2011)
CD3ε_{cyt} (Fernandez et al., 2010)

Figure 1. Mode of binding of intrinsically disordered proteins to lipid bilayers depends on the cell membrane model. The use of inappropriate models can result in physiologically irrelevant conclusions.

In 2000 (Aivazian and Stern 2000), using fluorescence, circular dichroic (CD) spectroscopy, and an *in vitro* kinase assay, the authors showed that the α-helical folding transition of ζ_{cyt} upon binding to acidic detergents and phospholipids (lysomyristoyl-phosphatidylglycerol (LMPG) micelles and dimyristoyl-phosphatidylglycerol (DMPG) vesicles, respectively) prevents ITAM phosphorylation. The authors concluded that this folding transition can represent a conformational switch to regulate TCR triggering (Aivazian and Stern 2000). Later, the formation of partial helices within the ζ_{cyt} ITAM motifs upon binding to LMPG micelles was analyzed using CD and multidimensional nuclear magnetic resonance (NMR) spectroscopy (Duchardt et al. 2007a).

In 2008, in the paper by Xu et al. (Xu et al. 2008), the authors extended the previously reported findings and mechanisms suggested for ζ_{cyt} (Aivazian and Stern 2000, Duchardt et al. 2007a) to CD3ε and addressed a question concerning the possible role of membrane binding for CD3ε$_{cyt}$ in TCR-mediated cell activation. Using purified CD3ε$_{cyt}$ and membrane models such as DMPG vesicles and bicelles composed of 1-palmitoyl-2-oleoyl-phosphatidylglycerol (POPG) and dihexanoyl-phosphatidylcholine (DHPC), the authors showed that CD3ε$_{cyt}$ binds to acidic phospholipid-formed bilayers and that this binding is mediated by electrostatic interactions between the N-terminal basic residues of CD3ε$_{cyt}$ and polar lipid groups (Xu et al. 2008). As shown by CD and NMR spectroscopy, this binding was accompanied by the folding of the protein and the formation of partial helices within the ITAM region, leading to the burying of the two ITAM tyrosines into the hydrophobic area of the lipid bilayer such that these residues are inaccessible for phosphorylation by LCK in an *in vitro* kinase assay (Xu et al. 2008). Based on these findings and the results from the fluorescence energy resonance transfer (FRET) experiments performed in Jurkat cells, the authors concluded that TCR triggering *in vivo* is regulated by dynamic membrane binding of the CD3ε$_{cyt}$ ITAM motif and that "sequestration of key tyrosines into the lipid bilayers represents a previously unrecognized mechanism for control of receptor activation" (Xu et al. 2008). Thus, for both cytoplasmic domains, CD3ε$_{cyt}$ (Xu et al. 2008) and ζ_{cyt} (Aivazian and Stern 2000), lipid-dependent helical folding transition was suggested by the authors as a main regulator for TCR signaling *in vivo*, resulting in the development of a conformational model of T cell activation (Aivazian and Stern 2000, Kuhns and Davis 2008, Xu et al. 2008). In 2006, using a variety of biophysical techniques, we showed that the binding of ζ_{cyt}, CD3ε$_{cyt}$, and FcεRIγ$_{cyt}$ to POPG vesicles, which would be expected to be a better model to mimic the cell membrane than micelles,

DMPG vesicles, and likely bicelles, is not accompanied by a disorder-to-order transition (Sigalov et al. 2006). The proteins remain similarly disordered in free and lipid-bound forms, thus suggesting a principally different mode for their lipid binding activity: binding without folding (Figure 1). Considering also the importance of the clusters of N-terminal basic residues and the non-importance of the ITAM residues for binding of $CD3\varepsilon_{cyt}$ and ζ_{cyt} to acidic phospholipid bilayers (Sigalov and Hendricks 2009, Xu et al. 2008), these findings suggest that in POPG-bound proteins, the ITAM tyrosines can remain accessible to kinase phosphorylation, thus questioning the physiological relevance of the ITAM helical folding suggested for $CD3\varepsilon_{cyt}$ (Xu et al. 2008) and ζ_{cyt} (Aivazian and Stern 2000, Duchardt et al. 2007a). Interestingly, phosphorylation of ζ_{cyt} resulting in a fully phosphorylated protein with a net charge of -5.5, does not alter its random coil conformation and only slightly weakens but does not block its binding to acidic phospholipid bilayers (Sigalov et al. 2006). This shows that partitioning of ζ_{cyt} into POPG vesicles is driven mostly by the clusters of basic residues rather than the overall net charge. Thus, in our study (Sigalov et al. 2006), we investigated lipid binding of fully phosphorylated ζ_{cyt} and not the phosphorylation state of the lipid-bound protein. Therefore, the interpretation of our work by Xu et al. (Xu et al. 2008) that "... lipid binding only partially inhibited ITAM phosphorylation (Sigalov et al. 2006)" is incorrect. Moreover, the main results and conclusions (Sigalov et al. 2006) oppose the results, conclusions and interpretations reported by Aivazian and Stern (Aivazian and Stern 2000) and by Xu et al. (Xu et al. 2008). Nevertheless, our study (Sigalov et al. 2006) was largely misinterpreted and used in many papers by other authors (Call and Chou 2010, Dave 2009, Guy and Vignali 2009, Kuhns and Davis 2012, Lillemeier et al. 2010, Love and Hayes 2010, Nika et al. 2010, Sangani et al. 2009, Shi et al. 2013) to further support the conformational models of receptor triggering reported in (Aivazian and Stern 2000, Xu et al. 2008).

The major concern is that in the studies by Aivazian and Stern (Aivazian and Stern 2000) and Xu et al. (Xu et al. 2008), binding of $CD3\varepsilon_{cyt}$ and ζ_{cyt} to acidic phospholipid bilayers was assumed always to be coupled with protein folding and formation of helices within the ITAMs making their tyrosines inaccessible for kinase phosphorylation. However, this assumption is not consistent with our current understanding of lipid binding activity of intrinsically disordered cytoplasmic domains of TCR signaling subunits. First, binding which is mediated by electrostatic interactions between positively charged $CD3\varepsilon_{cyt}$ and ζ_{cyt} residues with negatively charged polar groups of acidic phospholipids is not necessarily coupled with folding, which is mediated by hydrophobic interactions between ITAMs and lipid tails (Sigalov et al. 2006, Sigalov and Hendricks 2009). Further, helical folding of the ζ_{cyt} ITAM motifs can be induced by principally different kinds of agents such as LMPG micelles and DMPG vesicles (Aivazian and Stern 2000), and trifluoroethanol (Laczko et al. 1998), indicating that electrostatic interactions are not necessary to induce helical structure formation. Similarly, for CD3ε_{cyt}, folding can be induced by LMPG micelles, DMPG vesicles and POPG/DHPC bicelles (Xu et al. 2008). Interestingly, based on the NMR data reported by Xu et al. (Xu et al. 2008), CD3ε_{cyt} helical folding induced by LMPG micelles structurally differs from that induced by POPG/DHPC bicelles, suggesting that the structural features of folded CD3ε_{cyt} depend on the model lipid system. In addition, the authors reported that the CD3ε_{cyt} ITAM residues are not important for binding to acidic phospholipid bilayers (Xu et al. 2008). Thus, these observations collectively indicate that for CD3ε_{cyt} and ζ_{cyt}, binding and folding are mediated by different interactions (electrostatic and hydrophobic, respectively) and are not necessarily coupled.

Importantly, contradictory data from structural studies of lipid binding activity of immune signaling-related cytoplasmic IDRs (Aivazian and Stern 2000, Sigalov et al. 2006, Sigalov and Hendricks 2009, Xu et al. 2008) resulted in the open discussion (Fernandes et al. 2010, Gagnon et al. 2010) as to whether these IDRs fold upon binding to the cell membrane *in vivo*, thereby sequestering the tyrosine residues present in the ITAMs of TCR signaling subunits and thus making them inaccessible to tyrosine kinases that would otherwise drive signaling as was suggested for CD3ε_{cyt} (Xu et al. 2008) and ζ_{cyt} (Aivazian and Stern 2000). Resolution of this contradiction is of both fundamental and clinical importance because if the "conformational switch" model of TCR triggering (Aivazian and Stern 2000, Kuhns and Davis 2008, Xu et al. 2008) is right, we could in principle, control TCR-mediated T cell activation by modulating the lipid binding and folding activity of cytoplasmic TCR-related IDRs. Our studies from 2009 and 2011 (Sigalov 2011, Sigalov and Hendricks 2009) provide a molecular explanation for these contradictions and discrepancies and suggests that not all detergents and/or lipid compounds can be used as appropriate models to study IDP binding to the cell membrane. We analyzed membrane binding of intrinsically disordered ζ_{cyt}, CD3ε_{cyt}, and FcRγ_{cyt}, and found that depending on the model compound used, there are two different modes of their lipid binding activity toward acidic phospholipids (Figure 1): 1) coupled binding and folding that is characteristic for micelles and those vesicles that are unstable upon protein binding (DMPG), and 2) binding without folding that is observed in the presence of stable vesicles (POPG). As we suggest, initially in both modes, clusters of basic amino acids in the regions outside ITAMs bind to polar heads of acidic phospholipids while the ITAM residues do not contribute to binding at this stage. Then, in micelles (mode I), hydrophobic interactions between ITAMs and detergent tails promote folding of ITAMs, thus making ITAM tyrosines inaccessible for kinases as it has been shown for the ζ_{cyt}/LMPG micelle system (Aivazian and Stern 2000, Duchardt et al. 2007a). In vesicles, depending on the bilayer stability, initial protein binding to the membrane may (mode I) or may not (mode II) induce vesicle fusion and rupture and promote formation of ITAM helixes, which are stabilized by hydrophobic interactions with lipid tails in ruptured bilayers (Figure 1). This demonstrates, for the first time, that the use of lipid vesicles of the same size and surface charge can result in opposite conclusions regarding the membrane-binding activity of proteins and its physiological relevance. This also underlines the importance of ensuring the integrity of model membranes upon protein binding, especially in studies of IDPs. Hydrophobic interaction-driven folding of ITAMs observed upon the interaction of immune signaling-related IDPs with α-helix promoters (Aivazian and Stern 2000, Duchardt et al. 2007b, Laczko et al. 1998, Xu et al. 2008) is likely to be physiologically irrelevant. Finally, in spite of the data supporting the coupled binding and folding concept for the interaction of the studied IDPs with DMPG bilayers (Aivazian and Stern 2000, Xu et al. 2008), the observed hydrophobic interaction-driven folding of ITAMs results from a disintegration of DMPG bilayers and is unlikely to be of physiological significance. In summary, I would like to highlight that the studies discussed clearly show the critical importance of choosing an appropriate membrane model in studies of protein-lipid interactions and how substantially our improved understanding of basic biochemical mechanisms underlying both fundamentally and clinically important processes such as receptor triggering depends upon critical evaluation of the data and observations accumulated to date.

DO IDPS DIMERIZE?

The existence of specific homo-interactions between IDPs (Figure 2) was first reported in 2004 (Sigalov et al. 2004) when using a variety of biophysical and biochemical techniques, the ITAM-containing members of immune signaling-related family of IDPs including ζ_{cyt}, $CD3\varepsilon_{cyt}$, $CD3\delta_{cyt}$, $CD3\gamma_{cyt}$, $Ig\alpha_{cyt}$, $Ig\beta_{cyt}$, and $FcR\gamma_{cyt}$, were all shown to form specific homodimers. Later, the ability of some other IDPs to homodimerize was independently confirmed (Aguado-Llera et al. 2012, Berko et al. 2005, Danielsson et al. 2008, Frost et al. 2015, Lanza et al. 2009, Pieprzyk et al. 2014, Simon et al. 2008, Singh et al. 2008) extending the phenomenon to different classes of IDPs and suggesting its physiological relevance. It should be noted though that in most of these studies, dimerization is accompanied by a mutual or "synergistic" folding of two IDP molecules at the interaction interface.

CD studies (Sigalov et al. 2004, Sigalov et al. 2006) demonstrate that in contrast to other dimeric IDPs (Danielsson et al. 2008, Lanza et al. 2009, Simon et al. 2008, Singh et al. 2008), immune signaling-related IDPs do not fold upon dimerization, thus revealing for the first time the existence of specific interactions between disordered protein molecules.

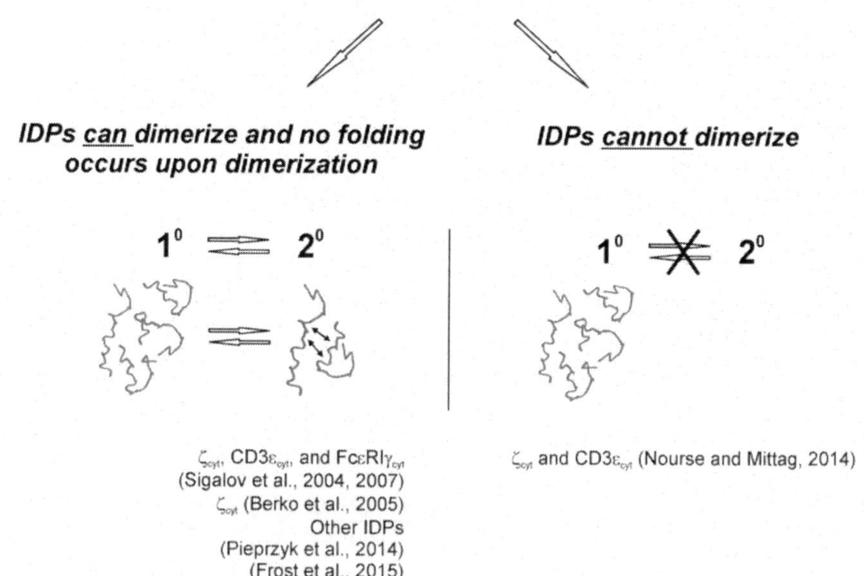

Figure 2. Contradictions in the field of dimerization of intrinsically disordered proteins.

Phosphorylation of ζ_{cyt} and $FcR\gamma_{cyt}$ does not alter their random-coil conformation in either monomeric or dimeric forms (Sigalov et al. 2004, Sigalov et al. 2006). NMR studies of ^{13}C-^{15}N-double-labeled ζ_{cyt} are in line with CD observations and strongly support the lack of a disorder-to-order structural transition upon dimerization (Sigalov et al. 2004, Sigalov et al. 2007). Interestingly, the lack of secondary structure in IDP homodimers was later independently confirmed for other unrelated IDPs: the *Aedes aegypti* N-terminal domain of Ultraspiracle -B (Pieprzyk et al. 2014) and the mammalian high mobility group protein AT-Hook 2 (HMGA2) (Frost et al. 2015).

However, despite the growing evidence that IDPs can dimerize (Aguado-Llera et al. 2012, Berko et al. 2005, Danielsson et al. 2008, Frost et al. 2015, Lanza et al. 2009, Pieprzyk et al. 2014, Sigalov et al. 2004, Simon et al. 2008, Singh et al. 2008) and that IDP dimerization is not necessarily accompanied by a disorder-to-order transition (Frost et al. 2015, Pieprzyk et al. 2014, Sigalov et al. 2004, Sigalov et al. 2007), the existence of this unusual biophysical phenomenon is still questioned and debated today not only in reviews (Sharma et al. 2015, Uversky and Dunker 2013) but also in structural studies of IDPs (Nourse and Mittag 2014). Intriguingly, in this study (Nourse and Mittag 2014), the authors studied the immune signaling-related IDPs ζ_{cyt} and $CD3\varepsilon_{cyt}$ similar to those studied in (Sigalov et al. 2004, Sigalov et al. 2007) but came to the diametrically opposite conclusion that these IDPs do not form dimers (Figure 2).

While a detailed analysis of potential reasons for this discrepancy is outside the scope of this piece, this contradiction highlights how unusual the biophysics of IDPs is (Sigalov 2010d, Sigalov 2011, Uversky 2013) and how important it is when interpreting biophysical studies of IDPs to consider each study's methodology, data quality, and validity. A specific example is the use of gel filtration or size exclusion chromatography (SEC) for characterizing the oligomeric state of immune signaling-related IDPs. In gel filtration studies by Sigalov et al. (Sigalov et al. 2004), the observed gradual shift in retention volume with increasing protein concentration, rather than the appearance of new peaks and a change in the intensity of peaks corresponding to monomer and dimer, indicates a fast dynamic equilibrium between monomeric and dimeric ζ_{cyt} and $CD3\varepsilon_{cyt}$ species. Further, cross-linking should theoretically trap the oligomers that are normally in a dynamic equilibrium with the monomers. In contrast to the gel filtration profile of non-cross-linked ζ_{cyt} at the same protein loading concentration, the profile of a cross-linked ζ_{cyt} sample exhibits separated peaks, with retention volumes corresponding to those of the monomeric and dimeric forms (Sigalov et al. 2004). The difference in gel filtration patterns for non-cross-linked and cross-linked ζ_{cyt} further supports the existence of a fast dynamic equilibrium between monomeric and dimeric ζ_{cyt} species in solution (Sigalov et al. 2004). Interestingly, when studying the oligomeric state of ζ_{cyt} by using a Shodex KW-803 gel-filtration column and the chromatography conditions similar to those that were used in our study (Sigalov et al. 2004): 20 mM potassium phosphate buffer (pH 7) and 1 mM EDTA, with 150 mM NaCl, the authors (Nourse and Mittag 2014) obtained data that are characteristic for dimeric ζ_{cyt} species (Sigalov et al. 2004). However, these data were compromised by combining the findings obtained using SEC conditions (0 or 1 M NaCl) that are inappropriate because of non-specific matrix-protein interactions (Nourse and Mittag 2014) and should be avoided in all SEC experiments in general (and especially when studying IDPs).

CONCLUSION

Two long-standing contradictions in the field of IDPs concerning dimerization and membrane-binding activities of these proteins that are briefly reviewed and commented here highlight the unusual biophysics of IDPs and the challenges of structural characterization of IDPs and their complexes. IDPs play an important role in cellular regulation, signaling and control in health and disease. For this reason, our expanded understanding of structure and function of can have multiple applications in biology and medicine.

The ability of IDPs to form specific homodimers and higher-order homooligomers (Sigalov et al. 2004, Sigalov 2011) represents a missing piece to the receptor triggering puzzle, and provides the molecular basis for the SCHOOL (Signaling Chain HOmoOLigomerization) platform of signal transduction (Sigalov 2006, Sigalov 2008b). In addition, according to this model, membrane binding of ζ_{cyt}, $CD3\varepsilon_{cyt}$, and $Fc\varepsilon RI\gamma_{cyt}$ does not affect phosphorylation of their ITAMs *per se* but prevents homooligomerization of these cytoplasmic domains in receptor clusters on the surface of resting cells and during random encounters of receptors diffusing in the cell membrane (Sigalov and Hendricks 2009).

Within the SCHOOL platform, specific protein-protein interactions in the cell membrane between the ligand-binding and the signaling subunits are critical for signal transduction and, as such, represent universal therapeutic points of intervention (Sigalov 2006, Sigalov 2010a). These transmembrane interactions can be specifically targeted by short synthetic peptides (SCHOOL peptides), which are designed to inhibit multichain receptors, in line with the SCHOOL platform based strategy, and can access their target site of action from both outside and inside the cell. Successful application of SCHOOL peptide technology (Sigalov 2008a, Sigalov 2010b, Sigalov 2014), both *in vitro* and *in vivo*, can stimulate the development of novel mechanism-based therapies. Thanks to the multiplicity and diversity of the receptors involved in the pathogenesis of numerous human diseases, the SCHOOL platform – together with the lessons learned from viral pathogenesis (Sigalov 2009, Sigalov 2010c) – can contribute significantly toward the improvement of existing therapies and the progression of novel therapeutic strategies for malignancies, thrombotic diseases, inflammatory diseases and diverse immune disorders.

REFERENCES

Aguado-Llera, D., Bacarizo, J., Gregorio-Teruel, L., Taberner, F. J. et al. 2012 "Biophysical characterization of the isolated C-terminal region of the transient receptor potential vanilloid 1." *FEBS Lett.* 586:1154-9.

Aivazian, D. A. and Stern, L. J. 2000 "Phosphorylation of T cell receptor zeta is regulated by a lipid dependent folding transition." *Nat. Struct. Biol.* 7: 1023-6.

Berko, D., Carmi, Y., Cafri, G., Ben-Zaken, S. et al. 2005 "Membrane-anchored beta 2-microglobulin stabilizes a highly receptive state of MHC class I molecules." *J. Immunol.* 174:2116-23.

Call, M. E. and Chou, J. J. 2010 "A view into the blind spot: solution NMR provides new insights into signal transduction across the lipid bilayer." *Structure* 18:1559-69.

Danielsson, J., Liljedahl, L., Barany-Wallje, E., Sonderby, P. et al. 2008 "The intrinsically disordered RNR inhibitor Sml1 is a dynamic dimer." *Biochemistry* 47:13428-37.

Dave, V. P. 2009 "Hierarchical role of CD3 chains in thymocyte development." *Immunol. Rev.* 232:22-33.

Duchardt, E., Sigalov, A. B., Aivazian, D., Stern, L. J. et al. 2007a "Structure Induction of the T-Cell Receptor zeta-Chain upon Lipid Binding Investigated by NMR Spectroscopy." *Chembiochem.* 8:820-7.

Duchardt, E., Sigalov, A. B., Aivazian, D., Stern, L. J. et al. 2007b "Structure induction of the T-cell receptor zeta-chain upon lipid binding investigated by NMR spectroscopy." *Chembiochem.* 8:820-7.

Dunker, A. K., Brown, C. J., Lawson, J. D., Iakoucheva, L. M. et al. 2002 "Intrinsic disorder and protein function." *Biochemistry* 41:6573-82.

Dunker, A. K., Silman, I., Uversky, V. N. and Sussman, J. L. 2008 "Function and structure of inherently disordered proteins." *Curr. Opin. Struct. Biol.* 18:756-64.

Dyson, H. J. and Wright, P. E. 2002 "Coupling of folding and binding for unstructured proteins." *Curr. Opin. Struct. Biol.* 12:54-60.

Fernandes, R. A., Yu, C., Carmo, A. M., Evans, E. J. et al. 2010 "What controls T cell receptor phosphorylation?" *Cell* 142:668-9.

Frost, L., Baez, M. A., Harrilal, C., Garabedian, A. et al. 2015 "The Dimerization State of the Mammalian High Mobility Group Protein AT-Hook 2 (HMGA2)." *PLoS One* 10:e0130478.

Gagnon, E., Xu, C., Yang, W., Chu, H. H. et al. 2010 "Response multilayered control of T cell receptor phosphorylation." *Cell* 142:669-71.

Guy, C. S. and Vignali, D. A. 2009 "Organization of proximal signal initiation at the TCR:CD3 complex." *Immunol. Rev.* 232:7-21.

Kuhns, M. S. and Davis, M. M. 2008 "The safety on the TCR trigger." *Cell* 135:594-6.

Kuhns, M. S. and Davis, M. M. 2012 "TCR Signaling Emerges from the Sum of Many Parts." *Front. Immunol.* 3:159.

Laczko, I., Hollosi, M., Vass, E., Hegedus, Z. et al. 1998 "Conformational effect of phosphorylation on T cell receptor/CD3 zeta-chain sequences." *Biochem. Biophys. Res. Commun.* 242:474-9.

Lanza, D. C., Silva, J. C., Assmann, E. M., Quaresma, A. J. et al. 2009 "Human FEZ1 has characteristics of a natively unfolded protein and dimerizes in solution." *Proteins* 74:104-21.

Lillemeier, B. F., Mortelmaier, M. A., Forstner, M. B., Huppa, J. B. et al. 2010 "TCR and Lat are expressed on separate protein islands on T cell membranes and concatenate during activation." *Nat. Immunol.* 11:90-6.

Love, P. E. and Hayes, S. M. 2010 "ITAM-mediated signaling by the T-cell antigen receptor." *Cold Spring Harb. Perspect. Biol.* 2:a002485.

Nika, K., Soldani, C., Salek, M., Paster, W. et al. 2010 "Constitutively active Lck kinase in T cells drives antigen receptor signal transduction." *Immunity* 32:766-77.

Nourse, A. and Mittag, T. 2014 "The cytoplasmic domain of the T-cell receptor zeta subunit does not form disordered dimers." *J. Mol. Biol.* 426: 62-70.

Pieprzyk, J., Zbela, A., Jakob, M., Ozyhar, A. et al. 2014 "Homodimerization propensity of the intrinsically disordered N-terminal domain of Ultraspiracle from Aedes aegypti." *Biochim. Biophys. Acta* 1844:1153-66.

Sangani, D., Venien-Bryan, C. and Harder, T. 2009 "Phosphotyrosine-dependent in vitro reconstitution of recombinant LAT-nucleated multiprotein signalling complexes on liposomes." *Mol. Membr. Biol.* 26: 159-70.

Sharma, R., Raduly, Z., Miskei, M. and Fuxreiter, M. 2015 "Fuzzy complexes: Specific binding without complete folding." *FEBS Lett.* 589:2533-42.

Shi, X., Bi, Y., Yang, W., Guo, X. et al. 2013 "Ca2+ regulates T-cell receptor activation by modulating the charge property of lipids." *Nature* 493:111-5.

Sigalov, A., Aivazian, D. and Stern, L. 2004 "Homooligomerization of the cytoplasmic domain of the T cell receptor z chain and of other proteins containing the immunoreceptor tyrosine-based activation motif." *Biochemistry* 43:2049-61.

Sigalov, A. B. 2006 "Immune cell signaling: a novel mechanistic model reveals new therapeutic targets." *Trends Pharmacol. Sci.* 27:518-24.

Sigalov, A. B. 2008a "Novel mechanistic concept of platelet inhibition." *Expert Opin. Ther. Targets* 12:677-92.

Sigalov, A. B. 2008b "SCHOOL model and new targeting strategies." *Adv. Exp. Med. Biol.* 640:268-311.

Sigalov, A. B. 2009 "Novel mechanistic insights into viral modulation of immune receptor signaling." *PLoS Pathog.* 5:e1000404.

Sigalov, A. B. 2010a "New therapeutic strategies targeting transmembrane signal transduction in the immune system." *Cell Adh. Migr.* 4:255-67.

Sigalov, A. B. 2010b "The SCHOOL of nature: III. From mechanistic understanding to novel therapies." *Self Nonself* 1:192-224.

Sigalov, A. B. 2010c "The SCHOOL of nature: IV. Learning from viruses." *Self Nonself* 1:282-98.

Sigalov, A. B. 2010d "Unusual biophysics of immune signaling-related intrinsically disordered proteins." *Self Nonself* 1:271-81.

Sigalov, A. B. 2011 "Uncoupled binding and folding of immune signaling-related intrinsically disordered proteins." *Prog. Biophys. Mol. Biol.* 106: 525-36.

Sigalov, A. B. 2014 "A novel ligand-independent peptide inhibitor of TREM-1 suppresses tumor growth in human lung cancer xenografts and prolongs survival of mice with lipopolysaccharide-induced septic shock." *Int. Immunopharmacol.* 21:208-19.

Sigalov, A. B., Aivazian, D. A., Uversky, V. N. and Stern, L. J. 2006 "Lipid-binding activity of intrinsically unstructured cytoplasmic domains of multichain immune recognition receptor signaling subunits." *Biochemistry* 45:15731-9.

Sigalov, A. B. and Hendricks, G. M. 2009 "Membrane binding mode of intrinsically disordered cytoplasmic domains of T cell receptor signaling subunits depends on lipid composition." *Biochem. Biophys. Res. Commun.* 389:388-93.

Sigalov, A. B., Zhuravleva, A. V. and Orekhov, V. Y. 2007 "Binding of intrinsically disordered proteins is not necessarily accompanied by a structural transition to a folded form." *Biochimie* 89:419-21.

Simon, S. M., Sousa, F. J., Mohana-Borges, R. and Walker, G. C. 2008 "Regulation of Escherichia coli SOS mutagenesis by dimeric intrinsically disordered umuD gene products." *Proc. Natl. Acad. Sci. US* 105:1152-7.

Singh, V. K., Pacheco, I., Uversky, V. N., Smith, S. P. et al. 2008 "Intrinsically disordered human C/EBP homologous protein regulates biological activity of colon cancer cells during calcium stress." *J. Mol. Biol.* 380:313-26.

Tompa, P. 2005 "The interplay between structure and function in intrinsically unstructured proteins." *FEBS Lett.* 579:3346-54.

Uversky, V. N. 2013 "Unusual biophysics of intrinsically disordered proteins." *Biochim. Biophys. Acta* 1834:932-51.

Uversky, V. N. and Dunker, A. K. 2010 "Understanding protein non-folding." *Biochim. Biophys. Acta* 1804:1231-64.

Uversky, V. N. and Dunker, A. K. 2013 "The case for intrinsically disordered proteins playing contributory roles in molecular recognition without a stable 3D structure." *F1000 Biol. Rep.* 5:1.

Xu, C., Gagnon, E., Call, M. E., Schnell, J. R. et al. 2008 "Regulation of T cell receptor activation by dynamic membrane binding of the CD3epsilon cytoplasmic tyrosine-based motif." *Cell* 135:702-13.

INDEX